Scientific Literacy
and the Myth of the Scientific Method

Scientific Literacy and the Myth of the Scientific Method

HENRY H. BAUER

UNIVERSITY OF ILLINOIS PRESS
Urbana and Chicago

Illini Books edition, 1994
© 1992 by the Board of Trustees of the University of Illinois
Manufactured in the United States of America
3   4   5   C   P   5   4   3   2   1

This book is printed on acid-free paper.

Library of Congress Cataloging-in-Publication Data

Bauer, Henry H.
    Scientific literacy and the myth of the scientific method / Henry
H. Bauer
        p.   cm.
    Includes bibliographical references and index.
    ISBN 0-252-01856-7 (cl : acid-free paper). — ISBN 0-252-06436-4
(pbk : acid-free paper).
        1. Science—Methodology. 2. Science—Philosophy. 3. Science—
Study and teaching. 4. Occupational literacy. I. Title.
Q175.B25   1992
502.8—dc20   91-18627
                                                                    CIP

Contents

Preface

Ignorance and misconception can have regrettable consequences. Nowadays, science and technology are so inextricably a part of our lives that ignorance and misconception about them may have particularly regrettable consequences.

Ignorance of science is indeed widely and quite routinely deplored. That scientific illiteracy is prevalent has become a shibboleth. Schemes proliferate for more and better education in science—just as they did after the first Russian satellite orbited the Earth in 1957.

What has not been widely remarked, however, is the high order of prevailing misconception, as opposed to rank ignorance, about science and technology: misconception even or particularly among the very people who most loudly bemoan the prevalence of scientific illiteracy. Public pundits, not only the public, drastically misconceive what science is, what it means to be scientific, and what the relationship is between science and technology. Scientists themselves, moreover, and science writers hold many of the same misconceptions.

Perhaps the central fallacy is that there exists an entity called "science" about which sweeping generalizations can validly be made; for example, that science is characterized and defined by the scientific method (which, it is widely supposed, can be defined rigorously and unambiguously). In actuality, for most generalizations about science their opposite also carries some truth. Thus it is in a sense true and yet also a misconception that science prizes originality: the most original scientists usually have a devil of a time getting anyone to take their notions seriously when they first put them forward. It is in a sense true and yet also a serious misconception that scientists should publish all their data, all the facts they have gathered: without selection, publications would be meaningless or misleading. It is in a sense true and yet also a misconception that successful predictions prove the predicting theory to be right. It is in a sense true and yet also a misconception that science is just common sense; that science is reliable; that scientific goals can be attained more quickly as more resources are deployed.

These misconceptions bring with them a considerable burden of misguided public action over such consequential and expensive matters as defense, education, or the environment. Unburdening ourselves of mistaken notions about science will be profitable in practical ways, not only in intellectual ones.

But on what authority are these beliefs labeled "misconception"? Who knows better?

Over the past several decades, people in increasing numbers and from a variety of intellectual backgrounds have made it their principal aim to understand the roles that science and technology play in contemporary society. Some of these people are engineers or scientists seeking to influence policy-making and the administration of technical initiatives; others are scholars or practitioners of public administration or political science who have come to specialize in science policy; others again are historians or philosophers or sociologists of science who have recognized the need to go beyond their traditional disciplinary perspective if they are truly to understand what science is and what it means to human society. Though a genuinely unified community has not yet coalesced from these groups, at least an encompassing term is becoming standard: STS, which stands for science, technology, and society (or, increasingly, science and technology studies; in Britain, the label "science studies" is commonly used).

More widely shared than any label is the ambition to understand how science, technology, and other salient features of human culture influence one another. Traditionally, philosophy has concerned itself only with logic and method as the foundations of science; sociology, only with the interactions of scientists with one another and with other people. Scientists themselves, by and large, have concentrated on learning how to understand and manipulate nature, not on what justification there might be for calling what always works "true." STS, by contrast, seeks a holistic view in which, for example, the way scientists interact is acknowledged to influence the validity of the knowledge they gain. So STS is—at first, now—multidisciplinary or eclectic. And like any young subject, it is also primitive or naive, by comparison with the sophisticated, long-established disciplines of philosophy or physics. But in mitigation, many of the insights it offers are of practical as well as academic interest.

The purpose of this book is to illustrate how insights gained in STS can serve to help anyone—specialist or generalist, scientist or humanist, thinker or activist—be intelligent about the place of science and technology in modern life. In the first chapter, I show how misconceived have been attempts to measure scientific literacy, and how misleading the results. Chapter 2 demonstrates how misconceived is the popular notion of the scientific method, and chapter 3 describes how science actually works. Further misconceptions are detailed in chapter 4, and imperfections in the practice of science are discussed in chapter 5. Major consequences of our misconceptions are pointed to in chapter 6, and the final chapter seeks to put science into its cultural context.

In the interest of easier reading, I have not broken the text with citations or footnotes; instead, references and sources of quotes can be found in the Notes on Sources, which follows the text. Moreover, I have not tried to give chapter and verse for every idea or interpretation that could be traced to some other book or article. Where I do give references, they are intended more to assist in further reading than to establish authority for my assertions: this book is in the nature of an expository essay rather than a scholarly monograph, and I accept full responsibility for the opinions stated.

After the source notes, I make some suggestions for general further reading; those should at the same time be taken as acknowledgment of my intellectual debt to the authors mentioned, John Ziman in particular. I have also learned much over the last decade from my colleagues in the Center for the Study of Science in Society at Virginia Polytechnic Institute and State University, and from the many visitors to the center, and from the many conferences held there.

Particular thanks go to those who commented in detail on drafts of this book: Leroy Ellenberger, Joe Pitt, Michael Swords; and most especially Stephen Brush, who showed me how to make very substantial improvements. More personal gratitude goes to Barbara, for her tangible help and above all for my peace of mind.

1

Scientific Literacy

In 1957, when the first Russian satellite orbited the Earth, the reaction in the United States was that science education had better be improved lest the Soviet Union get ahead technologically and therefore militarily and economically also. Similar sentiments have been vented over the years almost without interruption. In the summer of 1986, editorials and articles commented on a new survey in such terms as "America's Scientific Illiterates." "Americans' Disdain of Science" described how science and scientists are portrayed in the media. In 1989 we read about "The Dismal State of Scientific Literacy" and that the "Past Decade Shows No Gain in U.S. Science Literacy." Sigma Xi, the long-established national society to promote scientific research, offered among its lectures for 1990 one asking, "Can American Schools Produce Scientifically Literate High School Graduates?" and another asking apocalyptically, "Can Democracy Survive Scientific Illiteracy?"

The level of scientific literacy is low, the pundits are all agreed; and, they are agreed, that is bound to have serious consequences for the whole society. To do something about this indubitable and critical problem, we could join in Project 2061, which aims to improve the scientific literacy of future generations; or we could respond to a plea for funds "to help stem the flow of irrationalism currently sweeping America. . . . The national level of scientific literacy is only 5%," we are told, and "36% of the population believe that astrology is 'sort of' or 'very' scientific. . . . Is the U.S. in danger of losing its technological and scientific leadership among the world community?" And why is it that "74% of teenagers believe in angels . . . [and] 50% believe in ESP"?

Now I agree that misconceptions about science are rampant. But they are rampant among scientists as well as humanists and social scientists, among science writers as well as the general public. They are rampant even among those who purport to measure or survey scientific literacy. There are things drastically wrong with almost everything that has been said about the supposedly critical state of scientific illiteracy. The definitions of scientific literacy are worse than inadequate; the measures of literacy discriminate against the most literate; and predictions of the consequences of scientific illiteracy are not supported by the evidence.

Definitions

The widely publicized surveys are based on the notion that scientific literacy has three components: (1) the substantive concepts within science; (2) the nature of scientific activity; and (3) the role of science in society and culture. So far, so good. Anyone who has a reasonable understanding of these three things might well be called "scientifically literate."

Scientists might prefer to add to the first component "facts" or "phenomena," but that would be largely a semantic quibble: there would be little disagreement over the actual content of the knowledge in question.

Trouble begins, though, when we look at how one goes about testing how good a person's understanding is of the three components.

Measures

As an indication of how familiar one is with scientific concepts, one is asked: "Do you have a clear understanding, or only a general sense, or little or no understanding of: molecule; DNA; radiation?" In the scoring, the person who has a clear understanding naturally emerges as more scientifically literate than someone who has only a general sense. Yet who can claim *with justification* to have a clear understanding of DNA? Certainly those who are most directly engaged in trying to unravel the intricacies of DNA's functions would hardly claim to have a clear understanding: they are aware, for one thing, that the function of much of the chromosomal DNA is not yet known. But those who know less about it might well imagine themselves to have a clear understanding just because they know that DNA has something to do with heredity.

That question (like many badly drawn multiple-choice questions) discriminates against the genuinely knowledgeable. Again, who could claim with justification to have a clear sense of radiation? How clear can one's understanding be when radiation needs in some cases to be described by equations for particles and in others by equations for waves? Would a truly insightful—or, dare I say, a truly literate—person claim a clear understanding of radiation, at least before physics has constructed a Grand Unified Theory?

Sophisticated students know that high scores come to those who give the *expected* answer, be it right, wrong, or somewhere in between. Most people would give the "right" answer to the question "Does the Earth go round the Sun, or does the Sun go round the Earth?" But the scientific literate might wish to say "Neither," or "It isn't known," or "That isn't a meaningful question," having in mind that the solar system seems not to be centrally located in the universe and that no absolute frame of reference within the universe has yet been found.

"Electrons are smaller than atoms." Of course—so long as one sticks to notions held in the first quarter of this century, when atoms and electrons were just little particles. But the person who has heard about quantum mechanical tunneling might not be so sure, knowing that electrons tunnel very much more easily than atoms because the electron's wave function spreads sensibly so much farther.

These quibbles over specific concepts pale, however, in comparison with one's dismay at how an understanding of "the scientific approach" is determined:

> To be classified as understanding the process of scientific study, a respondent had to report that scientific study involved (1) the advancement and potential falsification of generalizations and hypotheses, leading to the creation of theory, (2) the investigation of a subject with an open mind and a willingness to consider all evidence in determining results, or (3) the use of experimental or similar methods of controlled comparison or systematic observation. Responses that characterized scientific study as the accumulation of facts, the use of specific instruments (i.e., looking at things through a microscope), or as simply careful study, were coded as incorrect.

But historians (among others) have inescapably demonstrated that what actually happens in science cannot be described like that. The best one could say for that view of science is that mainstream philosophy of science and fledgling sociology of science were speaking in such terms fifty years ago. In later chapters, I shall look at ramifications of this wrongheaded view of science. For the moment, note merely that it cannot answer salient questions about science: What made the scientific revolution of the seventeenth century? If the scientific method produces reliable knowledge, why have scientific theories to be continually amended?

Respondents to these surveys of scientific literacy, then, are liable to be classed as literate if they hold superseded, inadequate views and as less literate, or even illiterate, if their understanding is more adequate. But, quite sensibly, the measurers do not rely on just this single question to decide whether one understands what scientific study is. Other questions are asked to determine whether people apply their understanding to actual cases. For instance, they are asked to characterize astrology as very scientific, sort of scientific, or not at all scientific, "because any person who correctly understood the scientific process would have rejected the notion of astrology as scientific."

Most people know, of course, that the official stance of science (and of social science, science writers, and others) is to declare astrology unscientific, indeed a pseudoscience, and so this question ought to offer little difficulty for most people. Puzzling, however, is the measurers' contention that those who properly understand scientific study *in the terms stated by them* (and quoted above) would therefore and thereby class astrology as not at all scientific. Perusal of the modern astrological literature—which does not, of course, mean the horoscopes published in daily newspapers—reveals at least a few attempts at the advancement of generalizations and hypotheses, leading to the creation of theory and subject to potential falsification through the mere fact that the hypotheses and generalizations have been advanced; and the astrological literature also displays attempts at systematic observation, sometimes associated with sophisticated statistics. Astrology is not at all scientific for the direct reason that the contemporary scientific community is agreed that it is not at all scientific, not because one can logically show that the practice of astrology (in all its variants) contradicts that simple formula of the scientific method.

The disparaging opinion of astrology within the scientific community has no single or simple basis. Indeed, most scientists have never bothered themselves about the matter. It seems to have happened that, by some centuries ago, virtually all scientists had realized that astrological technique was not getting anywhere, and so for a long time now very few scientists have paid much attention to it. Astrology is generally regarded as unscientific because the current paradigm can conceive no mechanism by which planetary positions could significantly influence human affairs, not because astrology necessarily abrogates the scientific method. That is illustrated by the case of Michel Gauquelin who uses "methods of controlled comparison" quite scrupulously; having found certain statistically significant correlations between achievement in sports and the position of Mars at the time of birth, he finds himself disbelieved and criticized *because the result is seen as impossible*, not because of failure in method.

What is actually being measured, then, by that question about astrology is whether one knows and accepts the prevailing consensus within the scientific community, not, as claimed, whether one understands what scientific study is and applies that to the issue at hand. So this question really tests again the first component of scientific literacy—substantive concepts of science—not the second; or perhaps it reflects the third component, a recognition of the role of science in society: when science says something is unscientific, it speaks with authority and is to be believed. In itself there is nothing much wrong with measuring those things, of course, if only it were done with conscious intent. But other questions reveal that, to answer in accordance with the prevailing scientific consensus, in other words, to receive high scores for scientific literacy, one must give mutually contradictory answers to certain queries.

"In the entire universe, it is likely that there are thousands of planets like our own on which life could have developed." Agree and you're literate; disagree and you're not.

It is true that most of those scientists who have taken the greatest interest in the question nowadays agree that the likelihood is high; yet there is by no means a consensus throughout the scientific community. Moreover, any agreement there might now be is quite fragile because it is based on an incomplete understanding of such crucial factors as how solar systems form. At any moment, a new discovery or a new idea could turn the conventional wisdom around.

Not so long ago the common belief was that some encounter of the young Sun with another large body had drawn away from the Sun material that then formed into the various planets. Such encounters would be rare, hence planetary systems were thought to be rare, and planetary systems like our own were thought to be very rare indeed. As to the probability of life developing on an Earth-like planet, guesses varied all over the place in the absence of useful, reliable, relevant information. One who disagreed with the statement posed above would have been quite scientifically literate just a few decades ago.

Nowadays, the common belief is that planets as well as suns are formed by accretion from clouds of dusty gas that are thought to be ubiquitous in the universe. Therefore planetary systems are now thought to be the rule rather than the exception, and among all those planetary systems there are likely to be a goodly number that include Earth-like planets. Furthermore, over the past several decades people have found that when they apply heat or electricity or radiation to mixtures of methane and hydrogen and the like (until fairly recently thought to be representative of the atmosphere of the young Earth), amino acids, the building blocks of proteins, are formed: ergo, life would naturally appear on an Earth-like planet.

(But it is an exceedingly far cry from those experiments to any believable scenario for the evolution of life on Earth. In fact, it is no longer universally accepted even that the Earth's atmosphere at the relevant time was rich in hydrogen or methane. So to be scientifically literate in this case means accepting a quite dubitable current view.)

If one does believe that life like our own is common in the universe—in other words, that extraterrestrial intelligence (ETI) exists—then it is natural to speculate about the possibility of making contact with it. The Earth is between four and five billion years old, but the universe is three or four times as old as that. Therefore Sun-like stars existed billions of years earlier than did our own Sun, and so Earth-like planets will have too. If the origin of life and its subsequent evolution are taken to be entirely natural (which is to say probable) events, at least on Earth-like planets some extraterrestrial civilizations will have existed for much longer than our own and will have had time to evolve far beyond our present state.

Even at our present stage, we see space travel and colonization to be feasible; and we know that communication far into the universe by radio or light signals or robot probes is unquestionably possible. Thus there should have existed for billions of years extraterrestrial civilizations with the technical means to make contact with us. *A scientific belief in ETI entails the likelihood that Earth has already been in contact with it,* albeit perhaps unwittingly. And yet, to be scientifically literate, according to these surveys, one must *disagree* with the statement "It is likely that some of the unidentified flying objects that have been reported are really space vehicles from other civilizations."

Now the *logical* link is tight between a high probability of ETI and a high probability of having already been in contact with it. So strong are the links in this chain of reasoning, in fact, that it is also used as an argument in the reverse direction. Since we have not yet been contacted, the argument goes, ETI does not exist. Perhaps Sun-like stars or Earth-like planets are not after all, for reasons not yet understood, as common as the conventional scientific wisdom leads us to calculate; or perhaps the origin of life is not so probable an event as the conventional scientific wisdom assumes; or perhaps evolution does not invariably follow so rapidly upon the origin of life as it seems to have done on Earth; or perhaps civilizations cannot but self-destruct soon after they attain something like our present stage of development; or perhaps truly advanced civilizations have no interest in exploration. Or, it can be argued, perhaps we *have* been contacted but do not realize it, because the extraterrestrials differ from us so drastically that we cannot recognize their communications, taking them instead as miracles or as mistaken observations or as paranormal phenomena.

No matter which of these answers one prefers, the inescapable point is that the conventional scientific wisdom does rate as so probable the existence of ETI that the dilemma is how to explain an apparent lack of contact. Under these circumstances, should it be regarded as unscientific, as scientifically illiterate, to believe that in point of fact there have been some contacts, and specifically that some of the *admittedly unidentified* objects that have been reported might be signs of such contact?

Clearly enough, these surveys of literacy require that respondents parrot the contemporary scientific majority view, even when that means subscribing to opinions that are logically incompatible with one another. Yet at the same time, respondents are supposed to view science as proceeding by a logically sound method!

To understand how the current scientific consensus can harbor mutually inconsistent views about ETI and UFOs, one must understand how science is both a social activity and an intellectual one. The prevailing scientific worldview is not a single, logically coherent entity so much as a mosaic of the beliefs of many specialized little scientific groups; and a belief gets incorporated in the mosaic if there is a scientific group espousing that belief. ETI, but not UFOs, is part of the contemporary scientific worldview because there exists an accredited tribe of scientists concerned with attempts to contact ETI (by means of radio signals or space probes), while there does not exist an accredited scientific tribe attempting to unravel the provenance of reports of UFOs.

Why should that be so? Perhaps in large part because of the manner in which research problems are chosen. Scientists choose projects that offer a reasonable prospect of success, for without being successful, scientists cannot make careers for themselves. The scientific search for ETI began only after it became technically possible to scan appreciable areas of the sky at the radio frequencies that such communications are thought most likely to use; but also and only after the conventional scientific wisdom had come to regard the existence of ETI as probable (for the reasons given earlier). So long as the scientific community regards the project as sensible, those who engage in it have a reasonable prospect of being able to publish their results, thus obtaining credit for their time and effort *even if they detect no signals*. Albeit negative, this would be seen as valuable information allowing hypotheses to be modified about likely frequencies, or about the nature of ETI, or about the probability of Earth-like planets.

By sharp contrast, the study of UFOs would lead to something publishable in the scientific press only in the event that definitive proof was obtained that UFOs are neither the result of human activity nor some already known type of natural phenomenon. Failing such a success, all time and effort and funds put into the work would be wasted from the viewpoint of one's scientific career. So the premium on getting positive results would be very high, at the same time as there exists no known way of definitely getting any results at all: UFOs are capricious phenomena, encountered only unexpectedly; they are not routinely observed on military or on airport radar, or by astronomical or satellite-observing telescopes, or by naked-eye search. Those who set out to study the phenomenon might never obtain a fresh piece of data, a lifetime could pass without a useful observation, and in the meantime there would be nothing to publish and no career. So on several occasions panels of scientists have concluded that nothing of scientific value is likely to be gained through ufology. Ufology is not a science because no way is known by which results of some sort or another, meaningfully negative or enlighteningly positive, could with reasonable probability be obtained.

That does not mean, however, that UFOs do not exist; nor does it mean that the scientific community should say that UFOs do not exist. Rather, science should simply tell the truth: that there seems to be nothing useful to be done about UFOs at this stage. That is not at all the same thing as saying that it is unlikely that some UFOs are extraterrestrial vehicles, and yet that is the statement evidently expected of scientifically literate people.

Achieving Scientific Literacy

That past surveys of scientific literacy may have been inept does not of itself signify that scientific literacy cannot be studied or achieved. But let us return to the definition of literacy and consider what it implies.

One could hardly quarrel with the view that a person should be called scientifically literate who understands the substantive concepts of science, the scientific approach, and the role of science in society. One might start to worry a little, though, as one considered how to lead people to such a state of literacy. To begin with, how many people do any of us know (or even know of) who might qualify?

Start with the substantive concepts of science. Obviously these must include the nature of the chemical elements, the compounds they can make (and how), and something of the behavior of those elements and compounds—at least such commonly encountered types as gases and liquids and solids, and why some solids are crystals and others are powders and yet others are glasses; and what polymers are and why some are fibrous and others not. Quite a bit of organic chemistry, because one needs that for biochemistry and physiology, and surely a scientifically literate individual would have at least a general sense of what is involved in nutrition and medication. But all that requires some elementary knowledge of physics, too—sound and light, electricity and magnetism; and something about elementary particles and forces, for we live in a nuclear age and must therefore know about nuclei.

With all that, one can then appreciate and learn at least the outline—essential for the scientifically literate person—of the story of the universe: big bang, evolution of stars and galaxies, formation of solar systems from supernoval debris. And withal one can then acquire a necessary smattering of geology, an outline history at least of the Earth, for that is so unavoidably intertwined with the story of biological evolution, which of course is familiar to all scientifically literate people, who, so far as biology is concerned, also know something about cells and organisms, the chief classes of plants and animals, and how they are related to and differ from one another; and of course about the processes of cellular reproduction, and of sexual reproduction, including mutation and recombination, and how DNA enables us to understand all that; and about the basics of developmental biology, and how nature and nurture interact to produce both similarity and variation.

Doubtless that quick survey has omitted some things that any scientifically literate person really ought to know, but no matter; even this rudimentary list surely makes the point. Are we really prepared to insist on this, recognizing that it calls for several years of college-level science? Those topics are covered only if one takes at least a couple of years of chemistry and biochemistry, and a year each of physics and geology and biology—college-level courses, that is, that would also presuppose a decent amount of high school science, not to speak of the mathematics and statistics needed for understanding physics and chemistry and biology. It is quite unrealistic to suggest that "all [college] graduates acquire a broad understanding of science and its impacts" by taking "15 to 16 credit hours of natural science." That is insufficient time to acquire a broad understanding of science, let alone its impacts; yet that is the sort of thing that blue-ribbon panels have suggested.

The ready answer to my objection, that so little science would not do the job, would be to say that some approach to broad understanding would be better than none. But this is by no means necessarily so.

Scientists are notoriously wary of teaching survey courses, that is, courses that range widely but do not penetrate deeply. The common view among scientists is that no one can appreciate science who has not actually *done* some science; hence the requirement that courses in science include laboratory or field work, and the common opinion that one becomes a scientist only after having done actual research and gained the Ph.D. degree—before that, one can be a technician but not a full-fledged professional.

There is considerable warrant for this view. In high school and in undergraduate courses in college, one learns from textbooks. One learns what is known with considerable certainty, in compressed form, so that one can remain quite ignorant of the fits and starts and false trails and silly beliefs that were amply displayed when that textbook knowledge was first being gained by human beings. Only when one tries to do something oneself—even if it is only the repetition of some standard experiment—does one begin to get a glimpse of what those equations in the texts actually stand for. Thus, having learned from lectures and books a host of chemical reactions involving organic substances, that oxidation of an alcohol yields an aldehyde and then an acid, for example, I discovered in the laboratory that when organic chemists talk about the product of a reaction they do not mean that the reaction proceeds cleanly, that the product is all that is produced. Far from it. Some reactions yield only a small percentage; and the product is the substance that interests the chemist, not necessarily the one that is produced in the most copious amounts. (Beginners usually find that tars are produced in the most copious amounts.)

A little learning of science may well be a dangerous thing. Particularly because textbook science is presented so dogmatically, many people who have had but a little science come to have too much faith in the facts and laws they have learned, and too much faith that science has all the answers. That unfounded degree of faith then turns out to be a hindrance as they try to make sense of public arguments. One has only to read of controversies over substances (radon, say) that are alleged, in trace amounts, to cause cancer after cumulative exposure, or about the alleged effects of a nuclear war (Nuclear Winter), or about almost any of the many controversies about technical matters, to recognize that proponents and opponents try to push their cases beyond what existing scientific knowledge can legitimately support and, at the same time, continually cite the authority of science for their view. One who believes that science embodies certain knowledge can only be confused as equally qualified experts invoke the sanction of science in opposing ways.

Technical controversies over matters of public importance invariably have to do with frontier science, not with textbook science. At the frontier of science, where research is pushing to acquire new understanding, there are ideas and hunches, probabilities and possibilities, and beliefs and supposed facts that often melt away as research continues. At the frontier there is new knowledge, but it is fragile, untested, fallible knowledge, by contrast to the long-established, well-tested knowledge of the textbooks. On any given issue, if the relevant knowledge is genuinely clear, if any undergraduate who has properly learned from the texts can produce the answers, then controversy does not flare. So if one wants a scientifically literate population in order that public policy about technical matters be argued by an informed electorate, then one wants scientific literates to understand the character of frontier science, and no amount of learning textbook science alone, the currently accepted concepts of science, would do the job. (But, of course, one cannot understand the content of frontier science without already knowing a great deal of textbook science.)

In any case, to acquire even a modest acquaintance with the substantive concepts of standard textbook science would call for training in science in much greater depth than our society seems prepared to enforce. Given that, perhaps we should examine again and more carefully just why we should want it. Is it knowledge about the natural world that is so important? But that is inaccessible—it is only knowledge of what the scientific community currently believes about the world that can be taught. Or might it not be even more useful to be aware of that, that there is a difference between knowledge about the world and knowledge of what scientists believe about the world?

As to the second component of scientific literacy, an understanding of what scientific activity is about, yes, that would surely be a good thing. But it would call less for the learning of science itself than for learning about the history and philosophy and sociology of science; that is, learning some STS. Though the claim is often made, especially by scientists, that one learns about science, about the scientific approach, about how to be scientific, through studying the content of science, all the evidence says otherwise. Through learning textbook science, one is misled about the nature of scientific activity by learning only about relatively successful science, the things that have remained within science up to the present. In scientific texts, one hardly ever encounters the phenomenon of unsuccessful science, and yet history teaches that the science being done at any given time will largely be discarded, even in the short space of a few years, as unsuccessful. Through learning from the scientific textbooks, one is liable to learn the misconceived view that science uses the scientific method, which is responsible for the success of science. Through learning from the textbooks of science, one learns a number of misconceptions about how science operates, as described in later chapters.

The third component of scientific literacy, understanding the role of science in society and culture, is unquestionably the most desirable one for citizens of a democracy. What remains to be established is how much actual science needs to be learned to that end.

Very little, it would seem, according to the surveys currently in vogue. The overall rate of scientific literacy in the United States in 1988, calculated by combining the measured literacy in each of the three components, was 5.6 percent. But understanding of the scientific process of thinking was at 12.1 percent, of scientific terms and concepts at 28.1 percent, and of the impact of science and technology at 49.9 percent.

In what matters most, then, appreciating the role of science and technology in culture and society, half the population was literate. More than a quarter was even literate in the language of science. Only 12 percent understood what scientific thinking is; but perhaps that low figure stems partly from the fact that the surveys used an outdated, unworkable definition of scientific study. Were that not so, and if the three supposed components of scientific literacy were not combined into an overall measure, there would seem to be much less cause for public concern.

What Would This Measured Scientific Literacy Be Good For?

Since the conventional wisdom is so misguided in its assessment of scientific literacy and so wrong in its teaching of what scientific activity involves, perhaps we ought to take another look at why the conventional wisdom believes scientific literacy to be so important. It turns out that those reasons will hardly stand up to sensible examination.

A cogent and concise summary has been given by W. M. Laetsch, who points out that scientific literacy is commonly argued to be desirable for the following reasons:

1. Knowing science enables people—as voters and as consumers—to make better decisions in what is a scientific age.
2. Understanding modern technology brings economic good and thereby aids national security.
3. Scientific knowledge supplants superstition.
4. If we learn to think scientifically, that is, in terms of the predictable consequences of actions, then our behavior will become more rational.

5. Familiarity with the scientific method will lead to a more ethical attitude.

The first two of these in particular seem to be accepted unquestioningly; yet they are as contradicted by the evidence as are the other three.

The first claim is discredited at once by considering the actual behavior of the most qualified scientists. Robert Oppenheimer and Edward Teller disagreed as violently as two people ever could about whether or when or why to build or not to build nuclear weapons; at the same time, they were both about as superbly qualified in the relevant science as anyone could be, and moreover both were anything but narrow—indeed, both were as generally literate as well as scientifically literate as anyone we are likely to encounter. If the marvelously broad and deep knowledge of science possessed by Oppenheimer and Teller did not enable them to reach agreement on what should be done, what earthly use would be the scientific knowledge that most other people could hope to acquire?

All public arguments about purportedly scientific issues offer examples of the experts disagreeing over what should be done. As to more everyday matters, Laetsch puts it very nicely: "The problem is illustrated by the many people who believe they are scientific literates but also believe that margarine made of unsaturated fats has fewer calories than butter, or that nuts have fewer calories and are better for you than an equivalent amount of pizza." Or again, recall that the conventional scientific wisdom says that no benefits accrue from ingesting large amounts of vitamin C, whereas the opposite is held by Linus Pauling, winner of Nobel Prizes not only for science (chemistry) but also for sociopolitical affairs (peace). The point is that no amount of knowledge of or about science in itself causes individuals or groups to make good decisions about the many quandaries of life: humans readily subjugate their knowledge to their wishes, believing and doing what they want, all scientific facts and knowledge notwithstanding.

The second claim is usually supported by the assertion that science spawns technology, an assertion that cannot be sustained, as will be argued later (chap. 6). For example, scientific literacy in the United States has been in acknowledged crisis for at least thirty years, and scientific education in the Soviet Union has long been acknowledged to be superior; why then is the Soviet Union still so technologically backward compared with the United States? And even though the quality of science in the United States has been unexcelled in the past half-century or so, it is Japan that has excelled in technological practice. The relation between science and technology is anything but straightforward; any relationship between scientific literacy and technology remains altogether to be established.

As to eliminating superstition, one might ask how it comes about that many scientists regard religion as superstition while many others maintain a traditional religious faith? Or how to explain the fact that scientific creationists, regarded as unscientific cranks by the majority of the scientific community, count within their ranks people qualified by doctorates in science and engineering?

As to the fourth point, it would imply that scientists, having learned to think scientifically, behave more rationally than do scientific illiterates. I am not aware that such a difference has ever been the subject of study, and common experience offers little to establish its existence. Certainly faculty meetings of academic scientists reveal no obviously higher level of rational decision making or behavior than among other academics. Knowing the consequences of actions leads more often to suppression of that knowledge than to more sensible behavior: that most homicides from firearms in the United States occur between friends or relatives has not quieted those (are there no scientists among them?) who insist that weapons are needed to fight off burglars and other intruders; that being overweight is harmful has not led people to eat less, only to jog between meals (and many scientists do jog); and so on. The individual behavior of most human beings, including scientists, and the collective behavior of humankind are not directly or effectively governed by intellect. And that vitiates the last claim too: the notion that better knowledge of nature would lead to more ethical behavior.

Do We Want Education or Indoctrination?

It is altogether a mistake to argue that scientific literacy, or any other sort of literacy, is necessary because it brings directly certain desirable, concrete social consequences. Yet people and pressure groups of all sorts parrot that same fallacious claim: "There is a growing recognition in the industrialized world that scientific literacy is an important component of long-term economic growth and of effective citizenship."

Racial barriers in science must be eliminated because "more scientists and a science-literate work force will be needed to retain this country's leadership position." (Why not eliminate the barriers simply because it is the right thing to do? How did this country ever attain leadership if racial barriers would prevent leadership being retained?)

"Scientific literacy is as vital as language, historical, or cultural literacy. Those who master science have the potential to wield great power over those who do not. A democratic society may flounder unless all citizens understand the spirit, character, and values of the science that empowers so much of society."

"Groups opposed to the fluoridation of water were winning local referendums, 'primarily because the majority of those voting were not scientifically literate'." Really? Or was it because the evidence for the benefits of fluoridation was inconclusive?

"To lick the drug problem, we've got to have intense, daily education. . . . We should give drug education the same priority as arithmetic," according to a pharmacologist at the University of Maryland. Other professorial enthusiasts in other disciplines proclaim that the only salvation is through becoming "culturally literate," or agriculturally literate, or information literate. The National Association for Science, Technology, and Society holds conferences on technological literacy (the sixth one was in 1990). Tufts University has established an Environmental Literacy Institute.

All this propaganda for literacy of one sort or another comes from people who believe that everyone should share their particular views of what the most important knowledge is and what conclusions should be drawn from it; in other words, they want others to be indoctrinated. But what they seem to call for, and even think they are calling for, is education—that everyone should be literate about agriculture/drugs/science/ . . .

That is sound rhetorical strategy. Education is something against which no one would have the gall to argue. So we talk as though holding the same opinion as ourselves can be equated with being educated. That is not so, of course; but by propagandizing in this manner, we seek to make our opponents appear to be against education rather than against our specific views.

What education can and ought to be is too large a question to attempt to settle here, but it must be raised. Much good sense about it can be found in the writings of Richard Mitchell, who publishes *The Underground Grammarian*. For the present purpose, it suffices to be clear that no amount of knowledge or information by itself results in any foreordained action, individual or communal. The whole proper point of education is to help people learn to think for themselves; and the actions of free people who think for themselves are disparate and unpredictable. As a successful practitioner of public relations put it, "educating the public was not necessarily conducive to a result. . . . they could be the most 'educated' people in the world, and still you could lose."

Literate people disagree with one another over literary matters. Economically literate people—economists themselves—disagree over economic matters. Scientific literates disagree among themselves over the

desirability of human exploration of Mars, of building a Superconducting Super-Collider, and of many other proposed public projects of
a technical sort. Those who call for some type of literacy because it will
bring with it specific, direct benefits—technological competitiveness with
Japan, say—are simply mistaken in their hopes.

I do share the belief that literate citizens make for a better republic.
But I interpret "literate" to include awareness of the contradictions
within the canons of knowledge and of the vagaries of individual and
social behavior, and therefore having a basis for being analytic and
critical. A literate citizenry is likely to do many things better than an
illiterate one, but one can hardly foresee which things, or how better.
A literate polity might, for example, prefer inefficiency with freedom
to efficiency with servitude, bearing in mind that Mussolini did make
the trains run on time.

Literacy, scientific literacy included, should be encouraged because
it is a good thing, not because it is a necessary tool for something else.
And if the notion of literacy as a mere tool can be done away with, then
we could perhaps talk judiciously about literacy and education and in
terms of something besides perceived crises and ever new dangers.

What Should Everyone Learn about Science?

That the commonly cited benefits of scientific literacy are illusory
does not mean that it is not good to be scientifically literate. One who
knows nothing of the current scientific worldview is an ignoramus, just
as is one who knows nothing of human history, or one who has read
nothing of the great literature of the world. We are all, to a greater or
lesser extent, ignoramuses; yet we also live more or less satisfactory
lives, more or less despite the ignorance. We can reasonably say that
less ignorance can make for more satisfaction, that the unexamined life
is not worth living, and that examination can be more insightful the
more we know of—among other things—science. We can reasonably say
that an acquaintance with science can lead to a richer existence. We
can legitimately say that scientific literacy is desirable. But we cannot
show that it is necessary, or more necessary than some other specialized
literacy.

Even the lesser claim of desirability, however, does not eliminate the
dilemma so long as scientific literacy continues to be defined in terms
that make its achievement impossible. But once the claimed economic,
technological, and social imperatives have been exposed as spurious, it
becomes a little easier to suggest what an attainably useful acquaintance
with science might be. It is clear that no bits of specific information

about one or more of the sciences could be nearly as meaningful as a sense of what position science and technology play, and have played, within human culture. That straightforward realization in itself would call for a complete revision of the conventional wisdom about liberal education: instead of making one or another science a required part of everyone's education in college, courses in STS ought to be a part of everyone's education. Of course, some science—and more than now—ought to be taught at school: children should learn not only about the birds, the bees, and the flowers but a little also about the galaxies and the stars, and about the substances that they encounter and use and of which they are themselves made. A great deal of that can be done within the framework of stories, more like natural history than molecular biology.

A nice, practical advantage accompanies the recognition that STS is more important than science itself for a liberal education. In learning science, each course builds slowly and inexorably on another, and it takes a long time before a perspective of meaningful insight can be attained; in the meantime, the small knowledge acquired remains a potentially dangerous thing. In STS, by contrast, just as in the humanities or the social sciences, a little knowledge is *not* so dangerous a thing. A little STS would be good, and more would simply be better.

In science, as we dig deeper, we need to unlearn what we learned before. We begin chemistry or physics not with wave mechanics and quantum mechanics but with particles and waves; we learn one approximation and use it for a while, until we are ready to abandon that for the next approximation, and then the next; and stopping at any stage but the last means that our view is simply wrong from the contemporarily accepted standpoint. That is, incidentally, what is wrong with so many pseudoscientific ventures: people try to build on superseded scientific knowledge, not understanding that or how it is superseded, and that their little knowledge has been for them a very dangerous thing.

In the humanities and in STS, our views become more sophisticated as we learn more, but we need not periodically shed one viewpoint to adopt the next. Take one of the staple understandings of STS, that facts are theory-laden (chap. 4). The notion is readily enough introduced through an analogy with vision, that humans see in certain diagrams only one of several possibilities—for example, either a rabbit or a duck (fig. 1). The first impact of this on students typically is the impression that there aren't any objective facts in science after all (or anywhere else, for that matter), that everything is a matter of interpretation, that knowledge is constructed, and that we cannot really discover anything about the world around us (if, indeed, there is a world around us). But further thought and observation reveal that some facts are indubitably more theory-laden than others, to a degree that matters quite profoundly: one can see a rabbit or a duck, but few people would see here a map or an extraterrestrial, let alone a typewriter or a screwdriver. That theory-ladenness is a matter of degree leaves open the possibility that there really is something objectively out there, even though our knowledge of it rests on interpretation. There are actually black lines and white spaces in that diagram, though there may not really be a duck or a rabbit. So the realism of scientists can be seen as legitimate even as we learn not to accept as necessarily real every item that is talked about by scientists as though it were real.

Figure 1. Is it a duck or a rabbit? Can there be a fact in the absence of a prior point of view?

We become steadily more sophisticated the more we learn of the history of science and about current scientific activity, and the more we examine in particular instances what the evidence indicates about a particular phenomenon and its explanation. Thus we can come to understand why scientists have enormous confidence in such things as the Periodic Table of the Elements and why that confidence is well placed even when the apparently equal confidence of some scientists in other things (cold fusion, say) may be obviously misplaced. It is not only sensible to think of scientific literacy as learning STS, it is eminently practical too.

The aim of this book, as already mentioned in the preface, is to show that we can think intelligently about science even if we know little about the substance of science—albeit we can enjoy a much richer life if we are able to appreciate what science has achieved and what extraordinary advances it is making.

2

The So-called Scientific Method

It is widely believed that the essence of science is its method. The earlier-mentioned definition used in surveys of scientific literacy expresses commonly held notions of what the scientific method is: systematic, controlled observation or experiment whose results lead to hypotheses, which are found valid or invalid through further work, leading to theories that are reliable because they were arrived at with initial open-mindedness and continual critical skepticism.

One universally acknowledged source of this view is Francis Bacon (1561–1626); knowledge, he said, should come by generalizing from what one actually observes in the world—by contrast with the classic, Aristotelian approach of deducing with logical rigor from axiomatic first principles. But over centuries of argument and refinement, it has become clear that Baconian, inductive work could never establish truly certain knowledge: the *next* observation might force a different view or theory to be adopted even if the previous million observations had not—for example, just because the first million swans observed were white, one could not guarantee that the next one would be white also.

Distinctions were then suggested between observable and nonobservable things, it being supposed that at least one could be certain about observables, even if knowledge about nonobservables was inherently less reliable. Karl Popper introduced the influential insight that theories could never be positively proven to be true, whereas some theories could sometimes be definitively disproved; so, he suggested, to be scientific meant to deal in theories that could—at least in principle—be falsified.

Certain other significant issues in the philosophy of science will be taken up later, for example, the "theory-ladenness of facts"—to what degree do we observe what we believe we shall observe, by contrast with what may (or may not) be really there? For our present purpose, it is sufficient to recognize that these are the salient acknowledged elements of the popular view of being scientifically methodical: empirical, pragmatic, open-minded, skeptical, sensitive to possibilities of falsifying; thereby establishing objective facts leading to hypotheses, to laws, to theories; and incessantly reaching out for new knowledge, new discoveries, new facts, and new theories.

The burden of the following will be how misleading this view—which I shall call "the myth of the scientific method"—is in many specific directions, how incapable it is of explaining what happens in science, how it is worse than useless as a guide to what society ought to do about science and technology.

Are Chemists Not Scientists?

The scientific method is empirical. Scientific theories result when observation confirms tentative hypotheses. When the evidence speaks against them, hypotheses are falsified and therefore discarded.

One of my fellow graduate students in chemistry at the University of Sydney many years ago was trying to calculate certain properties of molecules, and he was the first to try to take account of one relatively subtle factor. Unfortunately, his calculations turned out to differ from the experimental values by more than earlier calculations had.

According to the Method for being Scientific, Dave should have considered his calculations falsified and tried another tack. Instead, he and his faculty advisor *ignored the comparison with experiment!* They were both mighty pleased with the progress Dave had made. He graduated top of our class, not much later he was on the faculty at Oxford, and soon after that he was a Fellow of the Royal Society.

Dave is far from alone among chemists in trusting theory more than experiment. A few years ago, a review article in *Science* listed many instances in which calculations had been right while experiment had been wrong: for the energy required to break molecules of hydrogen into atoms; for the geometry and energy content of CH_2 (the unstable "molecule" in which two hydrogen atoms are linked to a carbon atom); for the energy required to replace the hydrogen atom in HF (hydrogen fluoride) by a different hydrogen atom; and for others as well. The author, H. F. Schaefer, argued that good calculations—in other words, theory—may quite often be more reliable than experiments . . .

That is the view of one who is a theoretician, of course. You do not have to be long in a chemistry department to learn that chemists are no homogeneous tribe but rather a (sometimes uneasy) confederation of several distinct tribes: the analytikers, the inorganikers, the organikers, and the physical chemists are almost universally recognized to be distinctly different; and among or within these, or occasionally as separate tribes in their own right, there are electrochemists, polymer chemists, theoretical chemists, and others as well. And there are further subdivisions still: for instance, within many of these tribes, into experimentalists and theorists.

Naturally, each tribe and subtribe thinks its own way of doing things to be the best way, the *scientific* way. So theorists tend to believe that experimental evidence is important only insofar as it suggests new theory; and if experiment and theory happen not to agree, the theorists will often prefer to believe the theory rather than the (experimental) evidence. Experimentalists, on the other hand, regard that as perverse; they know it is observation and experiment that teach us about how the world works, theories being only devices that make it easier to remember the facts.

Both sides have something of a point (though they rarely manage to get much beyond it). Taking for the moment the side of the theorists, it is unquestionably the case that failure to go beyond what experiment shows can mean that discoveries are missed. A dramatic instance is that of the structure of DNA (the molecules that convey hereditary information), whose elucidation is generally credited to James Watson and Francis Crick. One crucial bit of information was that DNA contains equal amounts of the substances A (adenine) and T (thymine), and also equal amounts of G (guanine) and C (cytosine). Those equalities had been indicated by the lengthy, painstaking experimental work of Erwin Chargaff, who has in more recent years made abundantly clear his belief that Watson, Crick, and the Nobel Prize Committee did not give him due credit. But when you look at Chargaff's publications, what you find are tables like those in figure 2, in which the amounts of A and T, and of G and C, are only *approximately* equal (or, what amounts to the same thing, in which the ratio A:G is only *about* the same as T:C). Chargaff wrote about his observations: "The results serve to disprove the tetranucleotide hypothesis. It is, however, noteworthy—whether this is more than accidental, cannot yet be said—that in all desoxypentose nucleic acids examined thus far the molar ratios of total purines to total pyrimidines, and also of adenine to thymine and of guanine to cytosine, were not far from 1."

By refusing to stick his neck out beyond the actual results and say plainly that they *mean* exact equality and hence some sort of pairing in the molecular structure, Chargaff may have missed out on a share in the Nobel Prize. Watson and Crick, on the other hand, were speculating and theorizing about the molecular nature and biological functions of

	In Ox Thymus			In Ox Spleen		In Human Sperm		In Human Thymus	In Avian Tubercle Bacilli	In Yeast	
	1	2	3	1	2	1	2			1	2
A(adenine)	0.26	0.28	0.30	0.25	0.26	0.29	0.27	0.28	0.12	0.24	0.30
G(guanine)	0.21	0.24	0.22	0.20	0.21	0.18	0.17	0.19	0.28	0.14	0.18
C(cytosine)	0.16	0.18	0.17	0.15	0.17	0.18	0.18	0.16	0.26	0.13	0.15
T(thymine)	0.25	0.24	0.25	0.24	0.24	0.31	0.30	0.28	0.11	0.25	0.29

	In Ox Thymus	In Ox Spleen	In Ox (average)	In Human Thymus	In Human Sperm	In Human Liver		In Human (average)	In Yeast
						Normal	Cancer		
A/G	1.3	1.2	1.29(±0.13)	1.5	1.6	1.5	1.5	1.56(±0.008)	1.72
T/C	1.4	1.5	1.43(±0.03)	1.8	1.7	1.8	1.8	1.75(±0.03)	1.9

Figure 2. Erwin Chargaff's data about the chemical composition of DNA, taken from his "Chemical Specificity of Nucleic Acids and Mechanism of Their Enzymatic Degradation," *Experientia* 6 (1950): 201–40.

DNA; and they postulated a structure in which the equalities are *exactly* 1—deviations found from that in actual practice could be regarded as experimental errors. Watson and Crick turned out to be (largely) right; so, once again, ideas or theory had turned out to be a better guide than raw data, to what it all means. Inevitably so, for raw uninterpreted data do not mean anything: meaning rests on interpretation.

Evidently, then, some of the most successful chemists have not practiced the proper scientific method, which is supposed to put evidence first and theorizing second.

Science is also supposed to seek new discoveries; but, it turns out, chemists often do not welcome new discoveries. For example, if you read about chemical reactions that oscillate periodically, you find that William C. Bray's discovery of such a reaction in 1921 was simply not believed. Some thirty years later, in 1951, a paper by B. P. Belousov on the same subject was rejected, the editor saying that the reported results were simply impossible. Finally in the 1970s these results came to be accepted, *but only after a theoretical treatment had shown how oscillations could come about.* Again, more heed had been paid to theory—which is to say to preconceived belief—than to plain empirical fact.

Is Anyone a Scientist?

I should make quite plain that I do not *really* want to say that chemists are not proper scientists. What I just did with chemists (because I happen to know them best, being one myself) could equally be done with astronomers, or biologists, or geologists, or physicists, or with any of the other tribes within science. The point I *do* wish to make is that purportedly authoritative pronouncements as well as popular ideas about how science works are very seriously mistaken. One can find innumerable examples in all the sciences where theory was believed in the face of apparent evidence to the contrary; one can even find such an approach explicitly defended by eminent scientists—for example, the physicist Sir Arthur Eddington: "it is also a good rule not to put overmuch confidence in the observational results that are put forward until they have been confirmed by theory."

Even worse, theory—which, remember, is preconceived belief—may cause scientists to think they are observing things that in actuality do not exist, like the canals of Mars. And all the sciences offer instances where major new discoveries have not been accepted for quite a while because they ran counter to existing beliefs—consider the discoveries of Hermann Helmholtz and Max Planck, of Joseph Lister in medicine, of Oliver Heaviside in mathematical physics, Thomas Young's wave theory of light, and the cases of Louis Pasteur and Gregor Mendel and Svante Arrhenius and on and on and on.

So the classical and common view of science misconceives the actual relationship between theories and facts; and (consequently, inevitably) it misconceives the nature of the scientific method—the things that scientists actually do. It misconceives the behavior of science and of scientists in the face of surprising discoveries; and it misconceives much else about science, about technology, and about their interaction with one another and with the wider society.

An important misconception is implicit in the very use of the terms "science," "scientists," "scientific." To talk of scientists is to imply that astronomers, biologists, chemists, geologists, and physicists are all somehow much the same in some significant respect. To talk of science is to imply that astronomy, biology, chemistry, geology, and physics are all much the same sort of things. When there is talk about being scientific, it is commonly implied that one can be that, scientifically methodical, irrespective of the particular nature of what is being done; that one can be scientific about anything, like canvassing for new members for a bridge club: "Westchester County . . . has a tremendous program . . . working on memberships scientifically—how to get the people and how to keep them."

As soon as one looks in any depth, however, it becomes less and less clear what is really the same about astronomy and biology, say, or about what astronomers do and what biologists do. Sure enough, both astronomy and biology (and the other sciences as well) have to do with the study of selected aspects of nature. Sure enough, their findings are always subject to the commands of reality: false results are discarded (sooner or later, as their falsity becomes sufficiently obvious). Sure enough, each of the sciences now offers impressively detailed, coherent, and reliable insights, far more than they did fifty years ago, vastly more than a century ago, almost unrecognizably more than two centuries ago.

But there, or about there, the identity among the sciences comes to an end. The diversity among them includes that they vary in the degree to which they use mathematics: physics and astronomy cannot do without high mathematics, whereas much of biology or geology needs little more than arithmetic, and various bits of chemistry fall into one or the other of those categories. Though this diversity is commonly acknowledged, not generally recognized is the degree to which that difference entails other significant differences of practice: whether or not quantification is seen as the ultimate aim, for example, and whether or not mathematics is a required part of a student's initial training, and whether or not one comes to equate "quantitative" with "scientific" or "good."

Again, the distinction between observational and experimental science is a commonplace, but appreciation of the corollaries is not. Yet the way in which observational astronomers work has little in common with the way experimental chemists work: they differ in the sorts of funding they apply for (telescope time or chunks of actual money), in their reliance on graduate students (optional as opposed to essential), in the frequency with which they are expected to publish, in the way their peers interpret the significance of articles published with many coauthors, and in all sorts of other ways as well. Observation is so much more at the mercy of nature than is experiment that meaningful distinctions are obscured when—as is so often done—the two are lumped artlessly together as alternative but somehow equivalent modes of being empirical.

Many other consequential distinctions among the sciences are less frequently remarked. For example, whereas astronomy and biology and geology are fundamentally and inherently concerned with large-scale change that seems always to have gone in the same direction, chemistry and physics are not. Astronomy has to deal with the evolution of the universe, the birth and development and death of stars; biology and geology seek to account for the evolution of living things and of the Earth. But physics and chemistry share no such concern with inherent, directional change: they delight, by contrast, in the discovery of permanent relationships, and they do experiments in which time is just another controllable factor. Again, astronomy and biology and geology are, by and large, observational sciences, studying whatever nature presents them with, whereas chemistry and physics, by and large experimental sciences, can decide what to study, within increasingly wide limits—to the extent of making materials and arranging conditions that nature never before knew.

With these and other differences among sciences come far-reaching differences in attitude and method on the part of those who do the science, differences not often explicitly recognized. For example, chemists and physicists do not mean quite the same thing when they call a thing "stable": physicists mean that it is stable for all time, that it is in its lowest state of energy and will remain there until disturbed by another object or a force, whereas chemists mean that the thing does not by or of itself change into something else *at a noticeable rate* in a normal environment. The differences among adepts of the various sciences go beyond matters of theory, method, and vocabulary to subtler habits of thought and even to customs of behavior, to such an extent that the differences among the sciences, not only between the sciences and the humanities, can aptly be described as *cultural* ones: they involve a great deal more than just knowing about separate and distinct aspects of nature. Thus biologists and experimental physicists use visual imagery more than do theoretical physicists. Much theoretical speculation and argumentation over very few facts is commonplace in paleoanthropology or in astronomy but not in chemistry or in geology. Physicists look to crucial experiments to decide among theories at one fell swoop, whereas astronomers are used to waiting for long periods of time for the accumulation of data to bring an end to the speculation. Nobel Prizes in physics have been awarded about twice as often for experimental novelties as for theoretical ones, but in chemistry, experimentalists have been so honored five or six times as often as have theorists. Eminent physicists were found to feel pressed for time more than were eminent biologists; and the physicists gave up research in favor of administration at an earlier age. In matters of politics, physicists are considerably more liberal, on the average, than other scientists. Rates of divorce were found to be three times as great among biologists as among physicists (some decades ago, one should perhaps stress, when all the rates were lower than they now are). Though there seems not to have been any systematic study made of the matter, illustrations such as these are readily enough found to make the point that sciences differ among one another along many dimensions and not merely in commanding "knowledge" about separate pieces of the natural world.

Once the point is recognized, reasons can readily be suggested for some of these variations. As science developed over the last few centuries, the growth of knowledge demanded specialization. But the specialization unavoidably and soon became much more than a concern with distinct sets of phenomena. Those who studied some things found that they progressed best by taking more note of theory, whereas others found themselves going astray if they ventured too far from observation—and so some specialties came to understand that experimental evidence should not be accepted until it has been confirmed by theory, whereas most sciences and most scientists at least claim to believe the opposite. Each science—and to a degree each specialization within each science—has thus come to be an idiosyncratic blend of theorizing and empiricism; and that brings inevitably with it distinct notions about what knowledge (in general!) is and about the degree to which knowledge can be said to be "certain." In turn, disparate views about the nature of knowledge lead to different judgments about what might be interesting, valuable, fruitful to study. What we call "science" nowadays encompasses a wide range not only of knowledge but also of diverse views about the nature of knowledge.

Consider, as a last illustration, geology and physics. Physics is very oriented toward theory: one learns physics as a set of mathematically formulated laws more than as a set of observed phenomena; theory serves as a substitute for individual facts. Most physicists asked about the attraction between two of the planets will look up their masses and do a calculation; it would not occur to them to seek direct observational data. Geology, by contrast, is taught primarily through description—of minerals, geographic and geophysical features, strata, and fossils; theory in geology is less specific than in physics and serves to explain after the fact and not as a substitute for individual facts. Naturally, then, physicists tend to regard quantitative theory as the epitome of science and of scientificity; and, secretly or not so secretly, they see geology and geologists as somewhat less than highly scientific. So, too, physicists have learned that it is possible to find distinct, single causes for the variety of phenomena with which they deal, the phenomena themselves being identifiably and distinctly discrete. And for these reasons, and also because they can control all the relevant factors, physicists know that they can perform "crucial experiments" that compel nature to deliver definite answers. Geologists, on the other hand, learn that their phenomena overlap one another, that diverse "causes" conjointly produce any given geological circumstance, and that the most scientific approach is not that of seeking crucial tests but that of "multiple working hypotheses," for in geology one must, over long periods of time, be willing to countenance the possibility that any one of several competing explanations may ultimately turn out to be the best.

In view of such differences, it should not be surprising, for example, that it was a physicist who pushed most dogmatically the view that the dinosaurs were killed off in a discrete event by the simple cause of the collision with Earth of an asteroid, the impact of which remains demonstrable through the layer of iridium deposited at that time at the boundary between the Cretaceous and the Tertiary strata. To paleontologists, however, that seems absurdly oversimplified. For them, the layer of iridium-rich material is rather or partly an absence of carbonate sedimentation from the oceans during long eras in which almost no limestone was formed; the extinction of the dinosaurs itself is not seen as an event but as part of the change, over the course of millions of years, in the number of species as well as of individual dinosaurs; and, moreover, that extinction is looked at as only one of a number of occasions within geologic time when the diversity of living species decreased markedly, making it a statistical fluctuation within the perpetual flux of the appearance and disappearance of species, not a discrete, unusual individual occurrence.

Thus geologists and physicists tend to approach even scientific problems in disparate ways. They learn differently what it is to be scientific, what the scientific method is; and so too do chemists and biologists and other scientists come to different and even contradictory views of what science is. Yet these characteristic differences are but little recognized, and the misconception remains widespread that there exists a single method whose utilization marks the whole of science. In point of fact, as just illustrated, there is not any single thing that one can usefully and globally call science; rather, there are many different sorts of science. Once one has said that science is the study of nature, and that scientific knowledge is valid only so long as it is not contradicted by nature, one has said essentially all that is truly common, without qualification, among all the sciences. Beyond that, one finds nothing but variation: in the degree of weight put upon evidence in comparison with theory, in the ease with which data can be gathered, and in innumerable other details.

Diverse Aspects of Science

Some of the ways in which diverse bits of science differ from one another are indicated in figure 3. In what follows, consequences of this diversity will be illustrated in an oversimplified way by comparing or contrasting whole sciences with one another; but these distinctions actually characterize most faithfully only quite small bits of science within any of the major disciplines. Thus chemistry as a whole is relatively mature, data rich, experimental, and data driven, but many bits of chemistry are not mature (the chemistry of metal clusters, for instance, or of catalysts); quantum chemistry is a recognized subdiscipline that is notably theory driven, not data driven; and so on.

Some of the variations among bits of science have to do with their individual degrees of maturity. Physics is as a whole the most mature of the major sciences, and it seems plausible that this is a cause underlying the fact that physics also has the greatest degree of unanimity within its ranks: hardly anyone within the discipline questions relativity or quantum mechanics, or that the salient task for the discipline is to achieve the theoretical Grand Unification of the Four Forces. There is little ferment over what should be taught in physics courses, or how. The professional journals stand in well-recognized differentiation of function and hierarchy of status. The professional societies work quietly and without fuss (except, occasionally, over matters external to physics itself, like what to do about the civil rights of physicists in the Soviet Union). By contrast, the young computer science is in ferment over

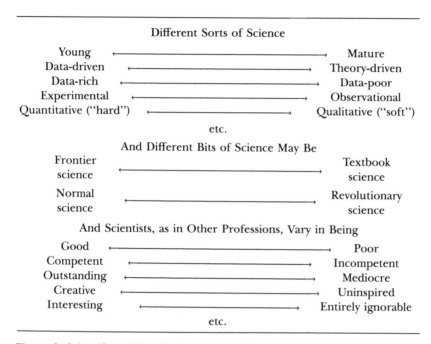

Figure 3. Scientific activity displays a very wide range of characteristics.

most of those things: what the most pressing goals of the field are; whether the subject itself is more akin to engineering or to science; how its practitioners should be trained; what constitutes professional publication; and so forth. Physicists look to governments for their major funds, whereas computer scientists look to their own institutions and to the computer industry. Physicists display the patina of an established aristocracy, whereas computer scientists exhibit characteristics of the nouveau riche.

It should be noted, however, that the theories commanded by a mature science are not necessarily more final and true than those commanded by a young science, though the assured behavior of the practitioners of the mature might lead one to think so. Physics for centuries has been the most mature among the sciences; and physicists periodically lapse into the belief that all the important principles of their subject have been discovered and that what remains is only to fill in the details. That was the case, for instance, circa 1870; but it was followed by disillusionment and then some of the most revolutionary bits of science: radioactivity, revealing that atoms are not stable and indestructible; relativity, altering drastically the notions of time and space and gravitation and motion; quantum phenomena, and the paradoxical view that some things could behave at times like particles but at other times like waves. Some of the most fundamental theoretical principles were replaced. (Yet, it should be noted, the vast majority of *knowledge,* in contrast to *theories,* in physics remained intact. A piece of uranium ore will always cause a photographic plate to become exposed, even through its protective black paper covering, and that tells us something definite about the external world; though the terms in which we try to explain that, connecting it with other phenomena, are always subject to some degree of change. That burning means combining with oxygen and not the release of phlogiston was the central realization of the most significant revolution in chemistry; but an enormous amount of what chemists know about combustion—which substances will burn in the presence of each other—is the same now as it was before that revolution. Thus much of the knowledge of what happens in nature, together with a great deal of the explanation that ties those phenomena together, remains essentially unaffected by apparently revolutionary shifts in theory.)

Again, mature science is not necessarily data rich; desired data are not necessarily acquired easily. To a possibly (but by no means certainly) apocryphal chemist is attributed the statement that he could not be bothered wading through the literature because he could get any answer he wanted more quickly by doing the experiment. Though physics is a more mature science than is chemistry, physicists would not venture that jocularity—though they might well say that they do not look it up because they can calculate it. Experiments in much of chemistry are more readily performed than in most of physics.

Much of chemistry is indeed data rich: feasible experiments can deliver a wealth of results in little time, and the chemical literature brims with data about millions of substances, about their preparation and properties. Consequently, chemistry has developed its idiosyncratic judgment about what and how much a chemist ought to accomplish. On the one hand, chemists publish—and are expected by their peers to publish—more articles than, say, geologists; on the other hand, chemists derogate those who do nothing but "turn the crank"—who use a given technique or instrument to derive data from a succession of different substances or reactions—because that is so easy a thing to do. Again because data are generated in such massive amounts, chemistry was the leader in developing techniques for trying to control the exploding literature, through "abstracts" journals and then computerized on-line abstracts. And again because so much can be done, chemists desperately want to have many co-workers to push their research forward, and much about university departments of chemistry can be understood as flowing from the overriding need to recruit plenty of graduate students, "pairs of hands" to carry out the necessary experiments. Thus there is related the story of X, a certain graduate student in chemistry who had the temerity to leave town with the wife of his supervising professor, Y. And X had the further temerity, a year or so later, to ask Y whether he could return to complete the work for his degree. As Y later said to a colleague, "Of course I agreed. He's first rate, and as you know, good graduate students are hard to come by while wives are a dime a dozen." One would not be quite as ready to believe that story if told about an astronomer or a geologist, for many excellent careers in astronomy and in geology, but few in chemistry, have been made without reliance on the labor of co-workers.

In data-rich fields, theory has a certain down-to-earth quality: speculation is fettered by the ease with which it can be disproved. But in data-poor fields, extremely tenuous chains of speculation are indulged. Thus cosmologists are notorious for theorizing whose equivalent on other subjects would be dismissed as science fiction: they are free, for example, to publish about what things may have been like before the big bang that started everything we know about; astrophysicists and cosmologists accept as conceivable the interpretation that certain observations are of objects producing inconceivable amounts of energy by means of inconceivable processes; those searching for extraterrestrial intelligence have published copiously, in almost total absence of data and in complete absence of any direct data; paleoanthropologists construct whole new charts of human evolution whenever a new fossil is found. By contrast, geologists denied continental drift for decades, supported though it was by fitting coastlines and biogeographical distributions. Geologists are always faced with a complex richness of data that offers continuing challenges even to meaningful categorization, let alone explanation, and so geologists are used to waiting and waiting for explanatory schemes to be developed; there is no hurry for that, explanation is not (yet) so salient a part of geology, and they have plenty of useful and time-consuming things to do without indulging in grandiose theorizing.

Some scientists thus do a lot of speculating, whereas others do virtually none, and there is no warrant to call the one approach scientific and the other not. It is just the case that different aspects of nature yield to investigation at different rates and in different ways, and so scientists come to differ in all manner of things. Whenever a generalization is made about science or about scientists, disregarding thereby the fact that there are so many distinct sorts of science, misconception is promulgated. What is true or fruitful for a field that is mature, data rich, and relatively quantitative (thermodynamics, say) is scientific for that specialty even though it may be entirely inappropriate and therefore unscientific for a field that is young, descriptive, and data poor (some bits of planetary science, say).

Again, though we use the single word "science" for both, *textbook* science is a very far cry from *frontier* science. What is in the texts is reliable. It is relatively uncolored by the personalities of those who originally conceived it. It is generally agreed to by almost all the experts. It is unlikely to need to be altered in the future, and in that unlikely event the alteration will likely be of limited extent. By contrast, science at the frontier is very unreliable: today's discovery often turns out to-morrow to have been an error. Frontier science often bears the stamp of its discoverer's persona; and it is often disputed by other experts. Frontier science and textbook science are about as different from one another as any two things can be, within the bounds that both are guesses about the nature of the real world. Our failure to bear these differences in mind has drastic consequences (as illustrated in chap. 6).

Scientists Are Human

Finally, the common view of science as a unitary, monolithic enter-prise fails to recognize how varied are the people who do it. Scientists are supposedly trained to judiciousness, objectivity, patience, and care-ful experimentation and observation; scant attention has been paid to how the practice of science is influenced by the fact that scientists, like all other human beings, vary in ability, competence, dedication, and honesty.

Indeed, thinking of science as using the scientific method portrays science as an activity that is highly unnatural: human beings are not by nature objective, judicious, disinterested, skeptical; rather, human beings jump to conclusions on flimsy evidence and then defend their beliefs irrationally. The widely held myth of the scientific method is one reason that scientists are often stereotyped as cold, even inhuman. Con-sider, for instance, what a celebrated humanist educator had to say:

> . . . teaching is an art, not a science. It seems to me very dan-gerous to apply the . . . methods of science to human beings. . . . a "scientific" relationship between human beings is bound to be inadequate and perhaps distorted. . . . to be orderly in planning . . . and precise in . . . dealing with facts . . . does not make . . . teaching "scientific." . . . A "scientifically" brought up child would be a pitiable monster. A "scientific" marriage would be only a thin and crippled version of a true marriage. A "scientific" friend-

ship would be as cold as a chess problem. . . . Teaching is not like inducing a chemical reaction: it is much more like painting a picture or making a piece of music, or on a lower level like planting a garden or writing a friendly letter. You must throw your heart into it, you must realize that it cannot all be done by formulas."

Thus science, arguably the finest exemplar of human intellectual achievement, is made to appear at best a necessary evil. When science is pictured as so impersonal and ascetic an activity, how to understand that scientists *do* throw their hearts into their work, which also cannot and is not all done by formulas? The myth of the scientific method hinders recognition of the wonderful diversity of the sciences. It makes it impossible to understand the history of science and contemporary scientific activity, and it fosters the stereotype of the cold, inhuman, sometimes evil scientist.

Genesis of the Myth

The false view of science as a unity defined by the unitary scientific method was not perversely adopted holus-bolus in the teeth of the evidence. It is just a naive and now-superseded view that congealed, for quite understandable reasons, during the nineteenth century. Moreover, that it was once a plausible view entails that it can still be made to seem plausible, at least under some circumstances and if one emphasizes some things to the exclusion of others. The classical picture is wrong in nuance perhaps more than through and through. Nevertheless, those errors of nuance have portentous consequences.

The roots of modern science, it is widely agreed, largely lie somewhere and somehow in the seventeenth century in western Europe. Though one can also trace continuity in important respects from earlier times than that into contemporary science, nevertheless pre-seventeenth- and post-seventeenth-century science are unlike one another in a striking way. Galileo and Newton epitomize that revolutionary epoch after which science grew at an exponential pace. Trying to understand what happened and why, it was natural enough to note that the Copernican revolution put the Sun rather than the Earth at the center of things, *where it actually is;* so centuries, even millennia, of careful and increasingly accurate observation had, it would seem, made plain *what the facts are,* and that had displaced preconception (or theory) based on (religious) authority. Galileo looked through the telescope; Newton saw the apple fall. Modern science seemed to have begun when theory—belief—came to be based on evidence, not on tradition or revelation.

And modern science began with simple, quantitative relationships among a small number of entities—bodies or particles, and forces.

Chemistry followed physics—albeit a century or so later—in becoming scientific, when the true theory of burning (substances reacting with oxygen) replaced the false notion of phlogiston, again apparently as the result of careful and quantitative observation in well-chosen crucial experiments. Geology followed, as humanity was forced to accept the evidence of strata, erosion, sedimentation, and so forth. Finally, Darwin brought the realm of living nature within scientific understanding by immensely detailed observation that pointed to a straightforward mechanism whose action is capable of yielding the variety of known biological species. So it was natural to see science by hindsight as a systematic progression toward true knowledge about the world, a progress made possible by inducing knowledge from observation; and this view could be bolstered by citing such people as Francis Bacon, who had explicitly advocated this approach.

One can trace over many centuries the intellectual struggle between, on the one hand, those who thought belief should follow authority, a priori reason, revelation, and the like, and, on the other hand, those who—like Bacon—thought that observation, experience, and evidence should be decisive. By the nineteenth century it seemed reasonable enough to most thinkers to believe that science had made triumphant progress by subordinating theory to evidence, and that the same sort of progress could be made in *any* field or form of knowledge—psychology, say, or mediumistic spiritualism—just so long as the evidence was gathered objectively and the theory based faithfully on it. This was the grand age of science, when it seemed to the leading scholars of humanity that the sure road to understanding all things had finally been discovered in science and its Rosetta stone, the scientific method. This was the time when T. H. Huxley preached what he called "lay sermons" in praise of the Church Scientific, and such accomplished scientists as Sir William Crookes, Sir Oliver Lodge, and Alfred Russel Wallace joined in the Society for Psychical Research to bring spiritual matters equally to understanding by means of scientific study. Social Darwinism flourished, Marx made history into a science, and Freud too based his theory on his evidence as he elucidated the workings of human personality.

By and large, the popular view of science remains much the same as that arrived at a century ago. That science is a powerful and progressive path to certain knowledge has been underscored by the proliferation of technology and of high technology, and especially by the harnessing of atomic energy in the 1940s. The classic description of the scientific method—theory based on and decided by evidence—has, despite its now-obvious inadequacies, been taught to and accepted by successive generations including the present one. But circumstances have made the classical view demonstrably obsolete: for one thing, over the past hundred years science itself has changed in many consequential respects; for another, our understanding of the birth and early development of science has become much more realistic.

Classically, scientists have been seen as dedicated truth seekers rather than as people who happen to hold jobs in science much as they might in, or as an alternative to, retail business, manufacturing, banking, or the law. Classically, they have worked as individuals, not in teams. Classically, they have been seen as curious and knowledgeable about the whole of nature even as they actively have studied only a relatively limited area of it. In times past, all these things were by and large true. But nowadays these things are no longer quite so true, at least not for the vast majority of those who work as astronomers, biologists, chemists, and so forth. Specialists in one field are little better than sheer amateurs in most other parts of science, and many of them are not even particularly interested in other areas than their own. Careerism, conflict of interest, and group-think are about as common now in science as in medicine, say, or in engineering or technology-based corporations. Most scientists now have a job more than they have a vocation.

Classically, the vices of scientists were scientific virtue taken too far, an excessive single-minded zeal in truth seeking: Faust bartered his soul for knowledge; Frankenstein's urge for technical accomplishment was not matched by humane understanding; Gottlieb and Arrowsmith (in Sinclair Lewis's novel *Arrowsmith*) suffered because of their very disinterestedness and naivete about human failings. Nowadays, however, the publicly revealed vices of scientists are seen to be quite the same as those of other people, avarice and dishonesty in particular.

As historians delve deeply into details of the past, it becomes abundantly clear that the classical understanding of scientific activity is far from the whole story—so far, in fact, that it needs more than cosmetic modification. The Copernican revolution was no simple triumph of evidence over preconception, for the simple reason that it could not have been: at the time, there were no decisive technical or computational advantages to Copernicus's approach over the long-standing Ptolemaic one. The Copernican venture was significant in subtler but also more deeply far-reaching ways—for example, in daring to make an individual intellectual choice in the face of long-established authority. For a time and to the benefit of a number of people, resistance to Copernicanism from established religion was avoided by emphasizing that this was only a model, not a statement about what the case might actually be in the real world. Attention to historical detail also reveals that it is not so clear why Galileo was condemned by the Church: if for Copernicanism, why did the condemnation come so late, two decades after Galileo had adopted the heliocentric view? As to Newton, some historians have found difficulty in accommodating the view of Newton as the ultimately objective scientist, drawing theories scrupulously from the evidence, with the fact that most of Newton's thinking time was devoted to alchemical and biblical exegesis rather than to mathematics and mechanics, as well as the fact that he quarreled bitterly with others over questions of scientific priority—not to speak of his use of "fudge factors" to improve the appearance of success of his science.

The corpus of science at any stage always includes only what has, up until then, stood the test of time. We see nothing in it of the trial and error, backing and filling, dismantling and rearranging that actually took place in the past, be that centuries ago or just a few years ago. Only when we read the actual accounts written by early students of nature do we begin to realize how many errors and false starts there were that left no traces in modern scientific texts. One can give excellent, objective, rational grounds now for the science in the textbooks, but that does not mean that it was actually assembled in an impartial, rational, steady manner.

The inadequacy of the classical view is not widely enough known, and much public discussion is still, implicitly if not always explicitly, based on it. Our rate of scientific literacy, as we have seen, is even measured by counting as literate those who hold classically mistaken ideas about the scientific method. Most universities specify that all undergraduates must study science, by which is typically meant a year or so of *any* science—as though that could be a useful sample of what science is, let alone what role it plays in culture and society.

The Epitome of Science

So long as science is viewed as monolithic, founded on the scientific method, it is possible (and therefore irresistibly tempting) to label some sciences, or some bits of science, as "more scientific" than others, according to the degree to which the method has been or can be successfully deployed. Hence, in the classical epitome of science, quantification and mathematics rule the roost, because hypotheses can obviously be framed and tested with real precision only when numbers are used. Immediately, then, biology and geology come to be seen as somehow less scientific than chemistry, which in turn is less scientific than physics, the *wholeheartedly* scientific science. Physics becomes the epitome of science.

That easy judgment is also abetted by historical circumstances. The scientific revolution of the seventeenth century was most dramatic in mechanics and gravitation, and until quite recently most historians of science worked actually in the history of physics. And since the history of science provides the grist for the philosophers of science, they too could become further confirmed in the opinion that physics is the ultimately scientific science. The most publicized scientist of the twentieth century, Albert Einstein, was a physicist. When the Nobel Prizes are disbursed in Stockholm, that for physics comes first. And physics, too, has been given the credit for achieving atomic bombs and nuclear power.

Yet if it is granted, among the many possible definitions of science, that science most fundamentally and undeniably means the study of nature, then it is surely a misconception that physics is the epitome of science: why should one think it more scientific to study waves and electrons than rocks or polar bears? Nor is physics less error prone than other sciences: once-accepted beliefs (theories) have been modified or replaced in physics as in other fields of science—indeed, perhaps more dramatically and unexpectedly than in other sciences. It is incidentally also a misconception that atom bombs and nuclear power stand to the credit of physics: many physicists, it is true, worked at these projects, and Robert Oppenheimer, the scientific director at Los Alamos, had been trained as a physicist; but many chemists, engineers, mathematicians, and others worked on the project, and the basic phenomenon on which all else was founded, the fission of uranium, had been discovered and understood primarily by the chemists Otto Hahn, Ida Noddack, and Fritz Strassmann.

Admittedly, physics is the most fundamental of the sciences in the sense that one explicitly needs some physics to do astronomy or chemistry, just as one needs chemistry to do any of the other sciences. Physics is also in a real sense simpler than the other sciences: it deals with phenomena that lend themselves to description in terms of a very small number of forces and things, so that simple relationships emerge. But that physics is fundamental and simple does not in logic make it the epitome of science (though it does offer plausible reason why human beings should have been able to succeed at it before they could get very far in more complex areas of science).

The misconception that physics is the epitome of science has unfortunate corollaries. In public life, one seeming consequence is that presidential science advisors are almost routinely recruited from among the ranks of physicists—and, moreover, no eyebrows, let alone protesting voices, have been raised over that. In other spheres it would be protested as grossly inequitable, not to say undesirable and potentially deleterious. To sense the latent significance for science policy, consider the analogy of defense policy: imagine that the chairman of the Joint Chiefs of Staff were routinely recruited from the ranks of, say, the army. Immediately everyone would recognize the possible conflict of interest and the subtle ways in which one-sided opinion could become too influential. So too with science, though popular lack of understanding masks the fact. Advice about research budgets and priorities would vary—on average only, of course, and in degree only—if it came from a chemist rather than a physicist or a biologist or a geologist. Moreover, no individual can be predicted to be more or less satisfactorily scientific by virtue of having been trained in one specialty rather than another.

In 1989, in the cause of improved economic forecasting, a conference between economists and physicists was mounted, again marking physics as the epitome of science, as though training in physics equipped one to be insightful and wise about anything and everything. In fact, physics might well be the *worst* possible training, within the options available in science, for advising on matters of social policy. The simplicity of the things with which physics deals easily leads physicists to look for single and simple causes and cures, so that they may oversimplify such things as the extinction of dinosaurs, the design of Star Wars systems for defense against missiles, or politics as a whole. Because physics deals with simple relationships, physicists are trained to be simpleminded. That is illustrated, for instance, by the ease with which accomplished physicists have been fooled by purported psychics and mediums, notably in the latter part of the nineteenth century and again in recent decades.

Because physicists, like almost everybody else, still see physics as the epitome of science, they are likely to make recommendations for other sciences that would be appropriate for physics but not elsewhere. Because of their training and the environment in which they have worked, it must be more difficult for a physicist than for another scientist to question the wisdom of dedicating more than six billion dollars (more recently estimated as sixteen billion) to the Superconducting Super-Collider, the next step toward the Grand Unified Theory. Yet that sum of money, if distributed among the other sciences, would treat them to a quite unprecedented sense of luxury; and I wager with great confidence that if it were put to the vote in a scientific senate with equal representation from each of the sciences, excellent other uses for that money would be found. Because the scientific method is universally thought to be universally applicable, recommendations appropriate to physics come to be not only made but also accepted about defense, education, and no doubt other human activities as well.

From Myth to Ideal

That the scientific method is a myth, that it does not explain the success of science and that scientists in practice do not follow the method, does not mean that the method itself should now be ignored or disparaged. Rather, it should be seen as an ideal—an admittedly unattainable ideal—not as a description of actual practice.

Those who hold ideals, no matter that they are unattainable, are likely to behave more in accord with them than will people who do not hold those ideals. Priests who vow chastity and poverty are likely to be more chaste and poor than people who do not make such vows—even though not every priest (and perhaps not even a single one) will be *entirely* chaste or *completely* poor. Politicians who believe that bribes are evil are less likely to accept them than are politicians who see nothing wrong with the practice. And so on. That human beings cannot by nature be entirely objective does not render objectivity an unworthy ideal: far from it, the ideal of objectivity in the form of disinterestedness, impartiality, or fairness is to be found not only in science but also in many aspects of social life. Those who strive to be objective in science can learn to be a bit more objective than they might otherwise be. Further, though it is well known that expertise does not readily transfer from one field to another, at least some of the people who have learned to be somewhat objective through doing science might be helped thereby to learn to be a little more objective in other matters too—just as some lawyers come to be fairly good judges of evidence and of human nature outside the courtroom as well as in it. There is no good reason to discard the scientific method as an ideal; rather, there is good reason to keep it so. Myths, after all, even if not literally true, are stories that embody moral truths.

That questing the grail is taught as an appropriate ideal does not mean that we should train undiscriminating Don Quixotes. It is important to make plain that the scientific method is an ideal, not actual practice. So long as the belief is widespread that science veritably follows the method, the belief will also be widespread, notably among scientists themselves, that *scientists* normally follow the method—if not perfectly, at least well enough. How then will the scientific community react when one of their number is found to have done some serious cheating? By regarding it as an individual aberration, almost certainly resulting from mental imbalance of some sort, an act that can have no significance for science as a whole, because scientists understand that they can make careers only by doing *objectively* sound work, for that is how they are judged, and anyone who tries to cut corners must have a screw loose. Through believing that the impersonal scientific method is actually in practice, scientists are thus kept from the realization that the progress of science—the progress, not the ultimate substantive content—is profoundly affected by the way in which scientists, as a community but also as individuals, behave. The greater the approach to objectivity, the better will peer review work; the greater the degree of honesty, the better will the whole system work; and, by contrast, the more corners are cut, the more students will imbibe the notion that corners are there to be cut and that the race goes to the swiftest, not to the best or the most sure. The myth of the scientific method keeps the scientific community from recognizing that they must have a humanly developed and enforced professional ethics because there is no impersonal method out there that automatically keeps science the way it ought to be.

The unqualified myth encourages hubris. One learns that science is objective. One learns that scientists are trained to be objective and to be skillful in use of the scientific method. Naturally, then, society learns to admire *scientists* as much as science itself, as people who are able to be objective; and scientists themselves, of course, are not immune to that chain of inference. So they may be led to think of themselves as more able to be impartial and free from conflict of interest than other people, like the Nobelist who was queried about the propriety of holding a university position while directing a commercially funded research institute: "I certainly do that to other people; I look at where they would benefit from a position and wonder whether what they're saying is a totally independent judgment. I think people are entitled to ask that of me. But I do think the statements and decisions I make come from the highest sense of integrity."

Instances are common enough in which successful scientists succumb to the temptation to see themselves as authorities not only in their own tiny field but over science as a whole and even beyond that; why not, after all, if application of the scientific method would make things so much better in all spheres of human activity? And because the public and the media also believe the myth of the method, great scientists are apt to be accepted as universal gurus (see chap. 4).

The myth of the scientific method, then, encourages the laity to have an unrealistic view of scientists and therefore also to have unrealistic expectations of them and of science; and it encourages scientists themselves to be unrealistic about themselves and about science, and to neglect the importance of cultivating consciously ethical behavior. It leads the scientific community to assume that its public credibility is permanent and quite automatically guaranteed—which makes it shocking and inexplicable when the credibility of science is brought into question, as in recent years under such well-publicized instances of misconduct as that of Thereza Imanishi-Kari, which was featured in congressional hearings and brought unwelcome notoriety also to her coauthors—notably Nobelist David Baltimore.

3

How Science Really Works

Science encompasses a wide range of particular fields that differ quite significantly from one another, in details of method and of epistemic belief as well as in content of knowledge. No generalization about the whole of science, about theory or method or fact, or about anything else, is valid across the board, without qualification. Scientists too differ from one another in many significant ways, in part because their different fields require it and in part because scientists vary as do all human beings.

On the whole, scientists share the belief that science is worthwhile, reliable, progressive. They share the belief that it is good to be scientific, even about matters outside science. They share the belief—the illusion, really—that the virtues of science, of being scientific, stem from using the so-called scientific method. That is to say, scientists by and large share the common, naive, and misconceived view of science; and further, scientists share the belief that they not only know what the scientific method is but actually use it.

Even though science is done very differently in the various specialties, and the consensus over how things ought to be done embodies significantly different emphases in the various fields, the different sciences nevertheless can be seen to have something in common if one focuses on the social activities that make up the enterprise of science.

Cooperative Action in Science: The Jigsaw Puzzle

Despite the drastic variations of consequential detail within science, there is an analogy that does instructively characterize the sort of activity that takes place in all the various domains of science. As Michael Polanyi has suggested, doing science is rather like putting together a jigsaw puzzle:

Suppose we share out the pieces of the jig-saw puzzle equally among the helpers and let each of them work on his lot separately. It is easy to see that this method, which would be quite appropriate to a number of women shelling peas, would be totally ineffectual in this case, since few of the pieces allocated to one particular assistant would be found to fit together. We could do a little better by providing duplicates of all the pieces to each helper separately, and eventually somehow bring together their several results. But even by this method the team would not much surpass the performance of a single individual at his best. The only way the assistants can effectively cooperate and surpass by far what any single one of them could do, is to let them work on putting the puzzle together in sight of the others, so that every time a piece of it is fitted in by one helper, all the others will immediately watch out for the next step that becomes possible in consequence. Under this system, each helper will act on his own initiative, by responding to the latest achievements of the others, and the completion of their joint task will be greatly accelerated. We have here in a nutshell the way in which a series of independent initiatives are organized to a joint achievement by mutually adjusting themselves at every successive stage to the situation created by all the others who are acting likewise.

Such self-coordination of independent initiatives leads to a joint result which is unpremeditated by any of those who bring it about. Their coordination is guided as by "an invisible hand" towards the joint discovery of a hidden system of things. Since its end-result is unknown, this kind of cooperation can only advance stepwise, and the total performance will be the best possible if each consecutive step is decided upon by the person most competent to do so. . . .

Any attempt to organize the group of helpers under a single authority would eliminate their independent initiatives and thus reduce their joint effectiveness to that of the single person directing them from the centre. It would . . . paralyse their cooperation.

Polanyi's metaphor, straightforward as it may seem, is capable without further ado of illuminating salient features of science: that modern science began when cooperation among scientists became widespread and systematic; that modern science is a quite particular sort of cooperative venture, working most successfully when autonomous; that what really constitutes pseudoscience is isolation from the scientific community; why science cannot be successful and also produce what ideologues want.

The metaphor can serve further to illustrate how the view came about that science is successful because of the scientific method. Early kibitzers—observers of jigsawing—will have noticed that errors were continually corrected and that players of the most diverse character and talents had been able to join in constructing the jigsaw puzzle. How could that happen? Obviously there must be a method, an impartially objective set of rules, that players follow to ensure that the moves each puzzler makes are impersonally objective and logical rather than reflective of that puzzler's personal ambitions, religious beliefs, gender, or what have you.

The puzzlers themselves found that explanation a good one (indeed, in earlier times there was no clear distinction between puzzler and kibitzer; both were avocations more than professions). At any rate, the idea of the method seemed able to explain the course of puzzling, especially after the fact. And it was agreeable, too, in placing not only puzzling but also puzzlers in a very favorable light, for the method incorporated some of the finest ideals for human behavior: disinterestedness, fairness, impartiality all flow from judging according to facts, not according to human prejudice or passion. Indeed, if only society as a whole would adopt the method, then everything would go better and everyone would be better off.

Thus kibitzers as well as players came to believe that the growth of their puzzle was owing to deployment of the method; but that belief, of course, does not make it so. The puzzle grows, provided there are participants who want strongly enough to make it grow. Over the past few centuries in particular, the puzzle *has* grown, and the picture it displays at any one time gives no hint of what had been there before or of how unmethodically individual players had actually conducted themselves.

Cooperative Action in Science: The Filter

Apt and handy as Polanyi's metaphor is even by itself, yet more features of science can be understood by examining in some detail how scientific communities generate scientific knowledge—in other words, what precisely the things are that puzzlers do.

Scientific knowledge—which aspires to be clearly expressed, with assumptions and limitations made explicit, reliable because well tested—can be pictured (fig. 4) as gleaned from a mess of all sorts of suggestions,

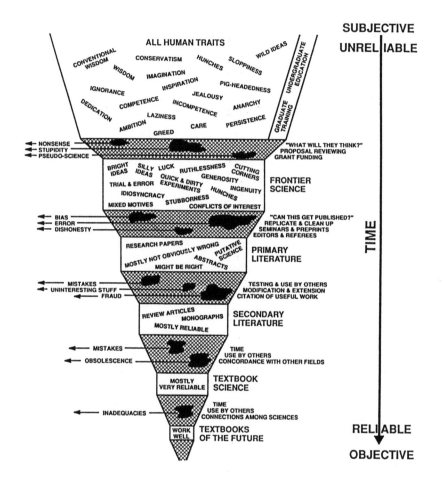

Figure 4. The Knowledge Filter. In stages, deficiencies are eliminated by virtue of the social institutions that science has evolved, peer review in particular.

claims, and beliefs by progressive refining as errors and inadequacies are filtered out.

Human knowledge begins with all sorts of suggestions from all sorts of people: just think of the claimed discoveries or facts of nature that are mentioned in the newspapers all the time—about psychic phenomena, alternative medicine, fluoridation of water, and so on. The world is full of beliefs, folklore, and wild ideas that compete with the conventional wisdom of the educated public and with expert scientific opinion. The world is full of people who push all sorts of beliefs and ideas with varying degrees of competence, persistence, and honesty.

Science is now—compared to a century or two or three ago—professionalized; it has evolved undergraduate and graduate training that are virtual necessities for those who seek to make valid scientific contributions. So what enters the knowledge filter of science is much constrained by what students learn before they become practicing scientists and generate new knowledge. One of the most important things they learn is to take account of their fellows in the scientific community, those who will review their proposals and decide whether and how much to support their research. Individuals therefore learn to curb or modify their wilder inclinations, to pay some heed to the consensus that prevails around them; and thereby much naivete and sheer nonsense are nipped in the bud almost before they can enter into science. Only those who have learned the current state of the art are taken seriously, and only when they comport themselves in reasonably disciplined fashion.

Research, or frontier science, is by no means a disciplined, homogeneous activity in which all the participants agree with one another about most things. Very far from it. Scientific research is a medley of all sorts of attempts to gain new knowledge, in every way that human ingenuity can conceive—by cutting corners; by doing "quick-and-dirty" experiments, not just carefully systematic ones; by following hunches or "just playing around," as well as by trying carefully thought-out things. After all, research is done by people who differ from one another in all the usual sorts of ways: in having noble or ignoble motives, great or little energy, many or few ideas, and so on.

The inchoate ferment that research scientists produce cannot become part of the accepted canon of science until it has been published; but getting published means convincing editors and referees that something about the work is sound and useful. And so, once again, individual frailties or imperfections must run the gauntlet of communal scrutiny, with the result that much of the error, bias, and dishonesty that exists within the ferment of frontier science does not enter the scientific literature. Scientists know that their data must look replicable and plausible and that their ideas about the data must be framed within already established knowledge. Published science is much more uniform in tone, format, and outlook than are the scientists whose names are associated with it; and that relative uniformity, together with the conventional use of third-person pronouns and verbs in the passive voice, lends the appearance of impersonality and objectivity to the so-called primary literature of published research papers.

But even published research is not yet scientific knowledge; as John Ziman points out, it is just information that has been made widely available. Unless it seems interesting to others, it will not be used and will fade from sight (and it is the case that the majority of the articles in the scientific literature are never cited by anyone). Those who do make use of a published piece of work thereby test it, and thereby also often modify or extend it. If they find something inadequate or wrong in it, they will make that known. So any piece of scientific work that becomes widely cited and therefore well known is unlikely to be fraudulent or to incorporate obvious mistakes; and dull, uninteresting, or scientifically incompetent stuff never becomes widely known. Only what has stood some test of time, as interesting and useful and not obviously wrong, becomes incorporated into the secondary literature of review articles, monographs, and graduate-level textbooks; and this then represents something like the prevailing consensus in the various research communities. It is pretty reliable stuff, mostly. But it is known in detail only to people who work actively in that particular field or in closely related ones.

The scientific knowledge that has widest currency is that contained in undergraduate textbooks, and there is an appreciable time-lag between something making its way into reviews or monographs and finding its way into textbooks. During that time, some things become obsolete and others turn out to be mistaken after all, so that what does get incorporated into the texts is mostly very reliable indeed. In this tertiary literature we find the textbook science that we learn through typically dogmatic teaching in school and undergraduate college and that we tend to believe unreservedly. Yet if we could look ahead, say, a hundred years, we would find the content of the textbooks different in significant ways: more coherent with other fields, less inadequate in a variety of ways, using viewpoints that are sometimes quite different (say, as particle ideas differ from wave ideas).

Overall, then, the raw stuff of frontier science has those characteristics of uncertainty, subjectivity, and lack of discipline that one should surely expect whenever human beings try to do what has never been done before. But after successive filterings through the institutions that science has evolved over the centuries, what remains easily gives the appearance of being objective and true. In point of fact, what remains is (relatively) impersonal rather than strictly objective, and it is hugely reliable and trustworthy rather than warranted true for all time; but in practice one rarely or never notices the difference—nor does that usually matter. John Ziman has ventured the guess that, in physics, textbook science may be about 90 percent right, whereas the primary literature is probably 90 percent wrong.

The Scientific Method versus the Filter

According to popular myth, scientific research is reliable because it is carried out methodically: hypotheses are precisely framed, crucial tests are envisioned and then objectively made. Any new bit of science could then claim much the same authority as any long-established bit; in fact, there would be no inherently significant contrast between textbook science and frontier science.

Only in the past few decades, with hindsight, has it become clear to informed scholars how plainly wrong that classical view is. One can never be entirely sure that any scientific discovery will continue to seem true for all time—as a Sidney Harris cartoon (fig. 5) so nicely illustrates. Even as we admire the reliability of scientific knowledge, we paradoxically take pride in the continual superseding of scientific views by more advanced ones—as another Harris cartoon (fig. 6) reminds us. Yet once-accepted but now-superseded scientific views were arrived at, as were the now-accepted ones, supposedly by exercise of the scientific method. Does that mean the method was applied incorrectly or inadequately in the past? In that case, would we claim that scientists now are able to be more objective and more precise in formulating hypotheses than were scientists in the past? Would we claim that even for such past scientists as Newton, whose views have also been superseded? Would we really want to claim that those lesser talents who stand on the shoulders of Newton and Darwin are better able to be scientific, are better scientists than those giants, even if they can now see farther?

Surely, we would rather admit that the scientific method, as a formula whose application can lead anyone to the truth, is a chimera. The progressive character of science is plausibly explained in the analogy of the jigsaw puzzle. The apparent objectivity of science results not from the cumulation of the individual objectivities of scientists but from the fact that the scientific community—the totality of puzzlers—works through consensus *because there is no other way to play effectively*. As Ziman puts

"WHAT'S MOST DEPRESSING IS THE REALIZATION THAT EVERYTHING WE BELIEVE WILL BE DISPROVED IN A FEW YEARS."

Figure 5. A Sidney Harris cartoon, reproduced with permission from *What's So Funny about Science?* (Los Altos, Calif.: William Kaufmann, 1977).

"IT MAY VERY WELL BRING ABOUT IMMORTALITY, BUT IT WILL TAKE FOREVER TO TEST IT."

Figure 6. A Sidney Harris cartoon, reproduced with permission from *What's So Funny about Science?* (Los Altos, Calif.: William Kaufmann, 1977).

it, science seeks to attain a consensus of rational opinion over the widest possible field.

Recognizing the mythical character of the scientific method has been a considerable relief to the philosophy of science (though, it must be admitted, not all philosophers of science have yet come to realize that and to benefit from it). For many decades philosophers had struggled to understand how method could logically explain the successes and the failures and the development of science; and after inductivism and positivism and hypothetico-deductivism and falsificationism, among other proposals, it became increasingly plain that there is no satisfactory explanation to be found along purely formal, intellectual, epistemic lines. Even if one could envisage a logically sound path along which science leads to accurate knowledge about the external world, one would still need to explain how that path at the same time perpetually leads to *inadequate* knowledge: that is, if the scientific method delivers results that require perpetual self-correction, then it is hardly a method that leads to certifiably reliable knowledge.

This realization, that more than philosophy is needed to account for the success of science, that the explanation cannot rest on purely intellectual grounds, is one of the reasons why some philosophers and historians and sociologists and others have come increasingly together in the scholarly interdisciplinary study (STS) of the activity that is science. One can comprehend science only by taking into account its history and its institutions, its social aspects as well as its cognitive ones; and the metaphors of puzzle and filter result from the determination to merge sociological with philosophical viewpoints.

In the following sections, the puzzle and filter model will be contrasted with the scientific method as modes of understanding such salient features of science as its diversity, its modern beginnings in the seventeenth century, and its relationship to pseudoscience.

The Scientific Revolution

Historians are agreed that modern science has its roots in the seventeenth century. Ideas, practices, traditions of research have been traced back in considerable detail without any need to regard nineteenth-century science, or eighteenth-century science, as a different kind of thing than modern science—distinct in many ways, to be sure, yet recognizably similar in approach to the study of nature. But continuity becomes much harder to discern as one goes back farther than that. One knows of bits of science in particular places at some times— in ancient China and ancient Mesopotamia, in classical Greece, in early

Islam—but there is not much coherent progression toward modern science to be discerned in them.

If modern science owes its success to application of the scientific method, then one has to regard the seventeenth century as a time when people, specifically in western Europe, first became really adept at drawing conclusions from observations, at testing hypotheses, at learning about the world from actual experience. Such an explanation is simply not tenable. From at least two or three millennia before that we have records of insightful discussions of empiricism and logic, and the history of Christian doctrine and biblical exegesis over nearly two millennia reflects continual tension between reliance on preconceived belief (or revelation, or theory) as opposed to empirical knowledge (or experience, or experiment, or evidence). Human beings knew about empiricism and skepticism and were capable of logic long before the seventeenth century, and not only in western Europe.

If modern science is recognized to be an inescapably cooperative, social activity, it becomes plain enough what was crucial in seventeenth-century western Europe: viable scientific societies were formed and scientific journals were established. Indeed, it is remarkable how well the measurable growth of those indicators of scientific activity over the last three centuries extrapolates back to a beginning in the seventeenth century (fig. 7). Until then, puzzlers played in relative isolation, all of them trying individually to put together all the pieces, like Polanyi's pea-shelling women; but in the seventeenth century, the puzzlers began to organize themselves, to specialize, to communicate rapidly with one another, and to act as critics for one another, to work on the puzzle together with increasing effectiveness.

Testing Hypotheses: Alone or Together?

One of the things wrong with the popular, classical definition of the scientific method is the implication that solitary people can successfully do good science, for example, frame hypotheses and test them. In practice, however, the people who put forward hypotheses are not usually the same people who apply the best tests to them. Proposers of theories do offer evidence for them, to be sure, evidence that in a sense tests their validity. But tests made by those who propose ideas could hardly ever be really decisive because scientists, like all human beings, are vulnerable to self-deception and have blind spots—most especially when it comes to things they are excited about and for which they have high hopes.

The most ballyhooed piece of putative new science in recent years, the claimed discovery of nuclear fusion in an electrochemical cell, offers

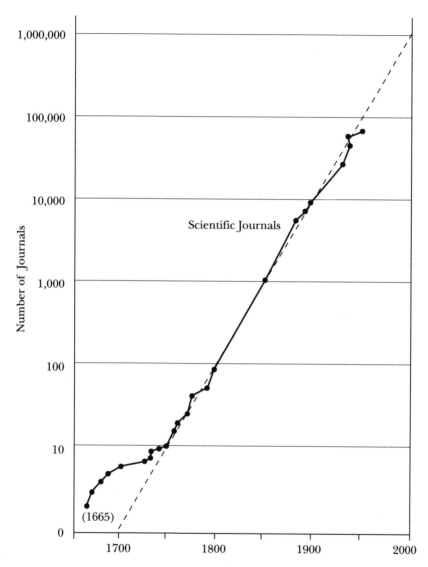

Figure 7. The size of the primary literature of science has grown exponentially, from essentially zero in the seventeenth century. Reproduced by permission of Yale University Press from Derek de Solla Price, *Science since Babylon* (New Haven: Yale University Press, 1975), p. 166.

a nice example. Among those who had up to then studied fusion (physicists trying to harness it, those who make hydrogen bombs, and those who explore the fusion reactions occurring inside the Sun and other stars), all expectations had long been that fusion of "heavy" hydrogen (which forms only one part in ten thousand of naturally occurring hydrogen) would occur more easily, by many orders of magnitude, than fusion of "light," or "ordinary," hydrogen. Now for several years two accomplished electrochemists worked to increase the amount of heat they could produce from palladium electrodes by what they believed to be fusion of heavy hydrogen; but they had not, immediately upon observing the production of heat by the supposed fusion, run perhaps the most obvious control experiments—namely, with ordinary (almost nonfusible) hydrogen in place of the (much more easily fusible) heavy stuff.

In all likelihood, they failed to conceive and perform this crucial experiment because, quite naturally, their focus and attention was on getting and testing ideas for improving the process they were convinced was occurring. After all, they had been led to do the experiments in the first place on the basis of speculation that fusion just might occur under the particular circumstances they chose. But no sooner had their claims been made public than others asked about such control experiments—again quite naturally, because those others were still thinking in accord with the belief that fusion cannot occur at all under the described conditions. The crucial test of the hypothesis had to be suggested by other people than those who got the idea in the first place.

There was, predictably, criticism of the claimants of fusion, and quite properly so: there is wide agreement that they should have done that control experiment before making a public claim. On the other hand, their failure to make that test was much more damaging to them than to anyone else, so the omission was hardly owing to malice aforethought. Most likely it was just the common human situation that people who have come to believe something find it virtually impossible to imagine that their belief might be wrong; and so they would also be unlikely to conceive experiments that could prove it wrong. Those scientists had spent some five years tinkering with ways of getting fusion, and they had been rewarded by some instances when remarkable amounts of energy were produced. They concentrated on increasing that energy, and on making the results more consistent. Specifically at what stage, one might ask their critics, should they then have done that control experiment?

Discoverers and kibitzers quite characteristically begin with opposing viewpoints: discoverers believe their discovery to be true even before the evidence is in, whereas kibitzers believe the conventional wisdom and will not accept a discordant or too surprising discovery until well after the evidence is in. Discoverers and kibitzers tend to interpret the same thing in opposite ways because they first look at it in opposite ways: the one group is attending to how full the glass is and the other to how empty it is, so that even when both are right (half empty equals half full), they disagree over what it means. Odd as it may seem to people who have not themselves been in such a situation, when one is trying to find ways of confirming a thing, one does not easily come up with ideas for proving that thing wrong, no matter how obvious to others such potential falsifications might seem to be.

The Diversity of Science

That the cooperative activity of science is recognizably similar in different specialties, even as those specialties differ in deeply cultural ways, also finds a ready analogy in the working of the puzzle. Though there is only one science because there is only one world, there are nevertheless many different parts to science, because each aspect of the world reveals itself best to a specifically evolved, idiosyncratic approach; and so the several sciences exhibit significantly different practices and there are many distinct sorts of science. Just like working the puzzle: though overall it is a single enterprise, it encompasses a large variety of areas and there are many disparate ways of playing. Some of the players will come to specialize in side-pieces and others will avoid them; and each group will come to regard the pieces with which it deals as more interesting than any others, and more valuable, and more indicative of how the work is progressing—as the epitome of puzzling, in other words. In some cases, players will find it most fruitful to pay attention only to the colors on the pieces and not to their shapes; whereas in other situations, color will be no help or even misleading while the shapes prove an excellent guide. After a time, some areas of the puzzle will contain just a few scattered clumps of pieces, and here the players may spend more time on theorizing than on actually adding new pieces; in other areas there will be so many obvious opportunities to add pieces that the players will defer until later any inclination to step back and get a perspective on what it all really adds up to, what it shows about this field and about the puzzle as a whole.

So different groups of puzzlers come to disagree about what the essence of puzzling is: theorizing or trial and error, color sensitivity or shape sensitivity, and so forth. And they develop different vocabularies to describe nuances of shape or of color. Thus do physicists differ from geologists over the primacy of theory or data, and from chemists over the meaning of "stable."

Kuhn's Model of Scientific Revolutions

Polanyi's analogy is consonant also with the influential description essayed by Thomas Kuhn, who pointed out that, most of the time, most scientists work at rather mundane problems of limited scope: that is, "normal" science. Certain things are measured to greater accuracy, mixtures of known things are studied, standard methods are applied to a variety of objects. So too with jigsaw puzzles: once the edges have been laid, and perhaps a few other clumps, the larger picture seems broadly apparent, and fitting the rest of the pieces in appears to be mundanely routine.

Every now and again, though, something unexpected happens—light turns out to behave not only like waves but also like particles, say—and then we have "revolutionary" science, or a Gestalt shift in some part of the puzzle. The pieces may turn out to have been quite deceptive, and what once seemed to be a nice fit must be partly dismantled and rearranged. Or what was thought to be sky with clouds, say, turns out to be reflection on the surface of water, and so there is a dramatic reinterpretation even as the bits of the picture remain unchanged. The players thereby can never claim absolute finality for any part of the picture, and yet as larger and larger areas are completed, it does become less and less likely that major Gestalt shifts or minor rearrangements will need to be made—the less likely, the more there are links to surrounding areas of the puzzle.

In normal science, consensus and conservatism and authority play a largely constructive role—approaches that have evolved as the most efficient ones continue to be so as long as the task remains much the same. Young Turks who insist that the clue of shape of the puzzle pieces should also be given weight, in areas where color has hitherto been a sufficient guide, find themselves ignored or resisted. But during revolutionary times, the old ways become a hindrance rather than a help: when progress has stopped, when some area of the puzzle seems unable to accommodate any of the available pieces, a rearrangement or Gestalt shift is needed, and that usually has to come from someone not yet entirely steeped in the old ways of doing things, someone who can envisage something never before seen. Scientists who shine during normal times may prove impotent as revolutionaries, just as those who lead revolutions might not have much to contribute once normal times return.

Still, science as a whole, knowledge about nature, has progressed through revolution and normality because of the value of conservatism and of innovation. Neither one is inherently more scientific than the other. Science needs many different contributions, and the scientific method is not just the testing of carefully constructed hypotheses. All sorts of method have combined to allow and to foster the growth and progress of science.

Kuhn's description of scientific activity rings true in many ways, and various bits of it have been adopted by natural and social scientists as well as by many philosophers and historians. But that description is not consonant with belief in the impersonal scientific method. Indeed, Kuhn's ideas were vigorously opposed in the first instance, especially by philosophers who saw them as making of science something fundamentally irrational rather than logical or rational. Such attacks still continue, but they come more from philosophers who stick closely to their own discipline than from those who engage with other viewpoints in STS, which recognizes science as an activity undertaken by fallible human beings, no matter how reliable the knowledge they have accumulated.

Science and Pseudoscience

A major, long-standing effort of the philosophy of science was to establish unambiguous criteria by which to distinguish science—the activity that produces reliable knowledge about the world—from pseudoscience, which misleads and falls inevitably into error. The scientific method does not provide such criteria; nor does it illuminate how scientists choose what problems to work at.

How could it be, some people have loudly asked, that science, which is always seeking new discovery, the more major the better, takes absolutely no interest in some of the most piquant possibilities? For example, that there may be relict hominids in the Himalayas (the yeti), the American Northwest (sasquatch or Bigfoot), and elsewhere; or that proper diet and mental health, supplemented by such simplicities as apricot pits, can guard one's body against cancer; or that UFOs evidence contact with advanced extraterrestrials; or that mindpower can bend spoons and see into the future as well as into the geographically remote present. On what basis can such claims be not only ignored but actually dismissed as pseudoscientific, as they typically are? If the scientific method exists, what excuse can there be for not using it to look at the most exciting possibilities that can be conceived? If other subjects can be studied scientifically, why not these? What prevents us from constructing hypotheses about these things and testing them, and thereby learning more about them?

The answer must begin, as already indicated, with the recognition that there is no such scientific method. It continues by noting that these "exciting possibilities" exist only in the undisciplined stew of suggestions that precedes even frontier science (recall fig. 4), which is even more disorderly and unreliable than is the ferment of frontier science. And, the answer can go on, it is also a misconception that science is always seeking major new discoveries: those almost always come unexpectedly, and moreover they are almost always resisted rather than welcomed, or at least resisted before they are welcomed.

"But it *shouldn't* be that way," one might well hear the complaint continuing from believers in the scientific method. "How can science progress if the conservative Establishment impedes progress like that?" Well, that depends on what sort of "progress" we're talking about. Remember what converts (some) frontier science eventually into relatively reliable textbook science: only the gauntlet of established scientific institutions. The price of reliability is deliberation in lieu of haste and conservatism in lieu of hitching rides on bandwagons. Individuals, of course, are free (in most societies) to rely on witch doctors, transcendental meditation, or laetrile, but science suggests that we stick to remedies whose effectiveness is consensually agreed upon by the international community of those who are professionally devoted to studying these things. A claim is not valid just because some people can tell us that they have tested and found good some hypothesis that they themselves constructed. Every quack can tell stories of success, and some quacks even believe their own stories and even have some grounds for doing so.

The salient point, again, is that the scientific method as classically formulated could obviously be applied by anyone to any investigation, and if it were application of the method that makes something scientific, then one could not label the study of anything "pseudoscientific" so long as the scientific method had been followed. If useful results happen not to be obtained, then the people concerned may have wasted their time, but one could still not label them "pseudoscientists."

In practice, of course, even those who use or preach the classical formulation of the scientific method in some contexts find themselves in other contexts invoking different (or additional) criteria for what counts as scientific: if, for example, reported results are too contradictory of established knowledge, then they are rejected even if they were apparently obtained methodically.

In a recent and notorious case, the journal *Nature* felt itself obliged to publish reports of reactions at inconceivably low dilutions of the reactants, results that would seem to support the possibility that homeopathic medicine might have a scientific basis after all, because no errors in method could be discerned by referees or editors. Yet the editor remained certain that the work could not be correct, and he was willing to publish the results only if the researchers agreed to have a team of investigators visit the laboratory and observe how the experiments were carried out. After that visit had taken place, controversy set in about the adequacy of the investigation; and *Nature* published a report claiming poor technique and likely deception or self-deception by some of the researchers.

It was the reported *results* that were unacceptable, but *Nature* felt obliged to search for inadequacies of *method*. A proper understanding of science could mitigate such dilemmas: a journal could choose to say that, even though no errors of technique had yet been demonstrated, because the results were quite unbelievable, there must have been something wrong in the experiments. That answer would have the benefit of being honest and might even lead the argument among editors, referees, and authors to become fruitful: by focusing on *why* the results seem so inconceivable, one might discover loopholes in the chain of reasoning and come to recognize the results as not entirely inconceivable after all. But even if there were no similarly happy outcome, such honesty would seem preferable to the confusion over ends and means that led *Nature*'s investigative team—composed of an editor, a magician, and a nonpracticing scientist!—to allege elementary errors of technique in the laboratory of an active, well-published—indeed, distinguished— scientist. It would be preferable if only because honesty is also supposed to be one of the norms of science that scientific journals try to uphold.

Uncritical believers in the scientific method find themselves in a bind not only in the face of such startling anomalies within the mainstream of science but also in the face of what most people might agree to be rank pseudoscience. Here are claims that they simply cannot believe, yet they are often reluctant to criticize the results as such because of the inescapable fact—which they admit—that science is never the last word: major new discoveries can come unexpectedly and therefore at any time. And all too often one can find in pseudoscientific claims no obvious transgressions of the method of observation, construction of hypotheses, and framing of tests. So, much too frequently, the uncritical scientific methodists allege methodic error or fraud without being able to substantiate the allegation. A notorious and well-documented instance is that of the Mars Effect. Michel Gauquelin found that champions at individual sports were born more frequently at times when Mars occupied certain positions in the sky, the frequency being upward of 21 percent compared to the expectation by pure chance of about 17 percent. During a long, heated argument, critics were unable to substantiate their allegations of deficiencies in method. Over and over again, the scientific methodists find themselves scratching for grounds to dismiss as unscientific investigations carried out with high competence and technical sophistication—for example, those in parapsychology in the Princeton Engineering Anomalies Research Program, or those at Loch Ness by the Academy of Applied Science.

If one understands that science is inescapably a cooperative enterprise, one can appropriately view as pseudoscience any claims made from outside the competent, relevant scientific community. Indeed, Martin Gardner used the term "hermit scientist" as a synonym for pseudoscientist, illustrating it with some well-known instances: Immanuel Velikovsky, who separated himself from the scientific community by refusing to accept the reviewing process for publishing in scientific journals, as well as by working almost entirely in isolation from everyday give-and-take; Ron Hubbard, who single-handedly invented dianetics, a complete system of psychology and psychological treatment; George McCready Price, who tried to make geology fit the Bible; Wilhelm Reich, who "discovered" orgone energy and spontaneous generation of life in the 1930s, whereas the scientific community had discarded in the nineteenth century the notion that life could arise spontaneously out of dust, dirt, or sand.

The social definition of pseudoscience also makes plain that there is a continuum from science to pseudoscience, not a simple dichotomy. The same sort of explanation then covers a whole range of eccentric episodes in and on the fringes of science. There are such cases as that of polywater, where a sizeable group of scientists in several countries fell into error; or that of N-rays, in which just a small, local community fooled itself; and innumerable instances of individuals bucking the conventional wisdom only to find themselves mistaken. (The famous examples of scientific heretics who turned out to be right are but a tiny number in proportion to all those who were heretical but wrong.)

Recognizing that one cannot make a sharp distinction between science and pseudoscience also accommodates the actuality that some subjects move from generally described as pseudoscience to being studied within science—acupuncture, say, or ball lightning: nothing has changed about the phenomena themselves, but they are now being scrutinized by a larger and more widely expert group of people. Similarly explicable are moves in the opposite direction, as with alchemy, astrology, or creationism: fewer and fewer scientists find such matters fruitful, and those who continue to study them drop out of the mainstream, getting farther and farther out as time goes by.

The scientifically methodical attempt to create two sharply distinct categories of science and pseudoscience also fails to account for the fact that the people who practice science are not always easily distinguished from those who practice pseudoscience. Thus Albert Szent-Gyorgyi and Wilhelm Reich would have seemed quite similar to anyone who met them and who was not professionally versed in their areas of discourse. Both were life-long mavericks, internationally peripatetic. Both were self-consciously committed to a positivist scientific ideal. Both were exceedingly charismatic and persistent, having to (and being able to) support themselves and their work largely outside regular scientific institutions. Why did one get a Nobel Prize whereas the other was judged a quack? The best explanation, I suggest, is because of their relative degrees of isolation from the mainstream scientific community. Szent-Gyorgyi maintained crucial links, most especially through submitting to the discipline of making his work publishable in conventional journals, whereas Reich took the easy way of founding his own periodicals, thereby foregoing the benefit of peer criticism.

Discarding the Myth of the Scientific Method

Under the scheme outlined in figure 4, scientific knowledge becomes, for good reasons, more widely valid and more firmly accepted as time goes by—those parts of it that are not jettisoned along the way, that is. At no stage, however, is guaranteed certainty reached; at no stage is it definitively proven that scientific knowledge is truly in accord with nature. Science is seen not to be dealing in permanent or absolute truth, as it was or could be seen if the scientific method could crucially test hypotheses against reality.

The myth of the method is not easily discarded, for one thing, because humankind is reluctant to accept that all knowledge contains an irreducible, inherent element of uncertainty. Over the last few centuries, the authority of science came to supersede that of religion precisely because science seemed to offer more certain knowledge, at least about the tangible world. If scientific knowledge now turns out to harbor ineradicable uncertainties, then science is in essence a false god, and, moreover, is inferior to the God on whom science turned its back. Human beings, after all, do want to be certain about fundamental things, and religion offers (or used to offer) such certainty. So there is reluctance to accept that the method is a myth, reluctance especially on the part of atheists, secular humanists, Marxists, and other such ideologues—perhaps the more so because fundamentalists and New Age obscurantists have also been so eager to topple science from its pedestal of authoritative certainty.

When Kuhn demonstrated unavoidably that science does not work logically and impersonally by means of the scientific method, he was quite bitterly attacked—on a number of specific details, to be sure, but chiefly because his views were interpreted by many people as undercutting the epistemic authority of science. If the only criterion of validity in science is the consensual agreement of the scientific community, it was said, then no safeguard exists against total error: whole communities, after all, have been known to harbor unrealistic beliefs. Without an impersonal scientific method, it was argued, science would be just like art or literary criticism—that is, at the mercy of emotion and fads.

These criticisms (which have by no means entirely died away) miss two essentials: first, the fact that the consensus of the scientific community governs the progress of science does not entail that this consensus is uninfluenced by nature; second, the fact is that history—including very recent history—offers ample instance where science *did* incorporate false belief, sometimes under the influence of emotion and fashion. To explain those instances, believers in the method have to say, "Well, the scientific method hasn't always been followed perfectly. But it *ought* to be. And if it were, everything would be all right."

Such an answer is hardly of interest to those of us who want to know about the role that science *actually* plays, and has played, in human culture. The fact is that science is not very like art or literary criticism, even though it has harbored seriously false belief at various times—and we should like to understand why. The fact is that science *as it actually exists* does not scrupulously use the scientific method, and therefore one cannot ascribe its success to that method. Yet science *has* been successful. The puzzle and filter metaphors afford plausible reasons for that success; the myth of the scientific method does not.

4

Other Fables about Science

Science, I have argued, comprises a very diverse set of activities that cannot be understood as just applying the formulaic scientific method, the popular myth notwithstanding. This myth is far from the only common misconception about science, however; there are a number of others that, in practice, foster a thoroughly wrongheaded attitude that is exemplified in unfortunate ways in public controversies. Some of these subsidiary misconceptions about science are described in this chapter, as illustrations. It is not the aim or the claim that this is a full tally of mistaken notions about science.

Science Deals in Facts

Science is commonly taken to connote fact or certainty. Thus the ubiquitous phrase in advertisements, "Scientific tests have shown. . . ." Leaving out "scientific" would not change the essential meaning—namely, that empirical evidence has confirmed the hypothesis—but it surely lessens the rhetorical impact. So too the popular quote from Isaac Newton, *"Hypotheses non fingo"* ("I feign no hypotheses"): when one deals only in observed facts, one has no need of hypotheses. Consider also the unqualified claim by many contemporary scientists and science writers that evolution is a fact, not a theory: in other words, when science speaks fact, no one has a right to doubt. One need only keep one's eyes open to find such usage in abundance, in scholarly as well as in popular writing.

Yet, as already remarked, no general claim to certainty made globally in the name of science can be sustained. Even those who hold "scientific" to be the highest accolade are wont to cite, as one of science's admirable qualities, that scientific theories change whenever the evidence requires it. The immediately obvious inference—which true believers in science fail to draw—is that every time there is a change, it indicates that something was unsatisfactory in the earlier version, in yesterday's science—which yesterday's true believers cited as the highest authority. Scientific theories are *beliefs* upon which explanations are based, and changes in theory are changes in belief, and such changes demonstrate that earlier beliefs were inadequate, unsatisfactory, or, not to mince words, *wrong*. Continually, some of yesterday's science is discarded as having been wrong.

Furthermore, scientific theory often does not even claim to be based in any obvious way on observed fact (see chap. 2). It may be held for no better reason than analogy with supposedly similar situations, as in the belief among physicists (up to the mid-1950s) that parity is conserved because mass, energy, and momentum are conserved; or it may be adopted for aesthetic reasons, through a preference for simplicity over complexity, or the like. One cannot validly maintain that in science observed fact always comes first while theory follows. Indeed, again as we have seen, science begins, in frontier science, under circumstances of great uncertainty; only after time and much indirect as well as direct testing do we get some textbook science in which there are to be found good approximations to facts in the sense of things that are certainly so. Overall, scientific activity in practice involves evidence of varying degrees of firmness and theories of varying degrees of plausibility. Something that is within science—said or written by one who is a scientist, or stated within a scientific organization or publication—is not thereby factual.

Pressed on this point, uncritical admirers of science have to concede, but they will typically qualify the concession by saying that admittedly not all, but certainly much, of science is so reliable that it would be perverse to contradict it. For instance, from an editor of *Science:* "Science can exist and is useful because much of the knowledge in it is more than 99.9% certain." But that sort of claim cannot be sustained either, unless the "much" can be specified for examination. As already suggested, one specification should be that we are dealing with textbook science, not with frontier stuff—which would, incidentally, already vitiate the usual aim of the true believers whose purpose is often to invoke the authority of science to discredit recent or contemporary claims as clearly or obviously wrong, say, about cold fusion or the existence of Loch Ness monsters. But even textbook science, of course, is not immune from the possibility of change; no discordant claim can legitimately be dismissed just because the textbooks seem to leave no room for it. One always comes down to the need to weigh probabilities *for the specific claim involved:* How good is the evidence for it? What is the likelihood that there is some undiscovered flaw in it? By comparison, what is the likelihood that now-accepted science can change in a way that could accommodate the seeming anomaly?

Once granted that probabilities are being compared, out-of-hand dismissal based on scientific authority is seen to be improper as a matter of principle. In practice, though, such dismissal is still attempted through various rhetorical devices; for instance, "Hypotheses with such small odds in their favor are usually said to be untenable." Those who want to use the authority of science to win arguments can easily find ways of insinuating that science deals in facts, since popular usage already fosters that illusion.

One of the modern recognitions by philosophers of science, and an axiom for students of STS, is that facts are theory-laden: that is, there is no such thing as a definite piece of indisputable knowledge about the world whose meaning is not in some way colored by preexisting belief about the world. The commonly used illustration of the duck-or-rabbit figure has already been mentioned (see fig. 1). To be noted for the present purpose is that whenever a "fact" is cited to prove a point, that fact has connotations that are less than absolutely certain but are inextricably attached to the fact by those who profess to believe it. A mountain is certainly a fact, not a theory, but no one (with the possible exception of the proverbial mountaineer) is interested in it just because it is there. A mountain is, variously and for different people, an illustration of plate tectonics, or a demonstration of God's handiwork, or an example of the incomprehensibility of the world, or the abode of certain spirits, or something else again; and each of these possible connotations determines how we see a mountain—as old or young or eternal, as generic or idiosyncratic, and so on. Whatever the characteristic involved, to each of us the mountain is something different; even though we all agree that it is a fact, it is *not quite the same* fact for each of us.

When evolution is said to be a fact, not a theory, what is actually meant? That now-living things have descended from ancestors, with modification, over time? Or that the modifications came by chance, not by design? Or, in addition, that all living things ultimately had the same ancestor? Or, still further, that the "first living thing" had as its ancestor a nonliving thing? Context indicates that when evolution is asserted to be a fact, not a theory, the view actually being pushed includes that of common origin, ultimate inorganic ancestry, and modification through nonpurposive mechanisms: a set of beliefs that goes far beyond the mountain of fact that is actually there, which consists largely of fossils that demonstrate *some* sort of relationship and *some* sort of change over time.

One consequence of misconceiving science as dealing in fact, or as capable of dealing in fact, is that science or scientists get criticized when they fail to deliver certainty or when they change their minds. The fundamentalist extremists who maintain the Earth to be only about 10,000 years old have committed this absurdity: science's current estimate of the Earth's age, they have suggested, is not to be relied upon because scientific estimates have changed so much over the years; why then should the latest estimate be thought trustworthy? Now as a point of principle that cannot be gainsaid. But the point is not the absolute reliability of the latest scientific estimate; rather, it is to compare the creationists' estimate with the scientific one. The latter was envisaged, in the nineteenth century, as hundreds of millions of years or more (except for the overly simplistic calculation of between 20 and 100 million years by the physicist Kelvin, toward the end of the century). For at least the past 80 years the Earth's age has been gauged in billions of years, largely consistent with evidence about the universe as a whole, as well as with information about the solar system as a whole, as well as with many indications of the Earth's own history. A rather coherent story, then, suggests billions of years; and that estimate has varied by no more than a factor of four over several decades; and the likelihood that it will change by as much as another factor of ten in the future, let alone by a factor of a hundred, is very small indeed. By contrast, the creationists have no direct evidence to support their assertion of about 10,000 years: all their arguments are negative ones, aimed principally at discrediting what science says.

The probability is clearly very high that the scientific view is more correct than the creationist one *on this particular point* for the sort of reasons just given. But that does not make it legitimate to assert that no one should take the creationist view. If a person wants to believe something that is exceedingly improbable, it would seem to be an inalienable human right to be permitted to do so, just so long as it does not harm others; and creationist belief hardly threatens to do that. (It must be noted, however, that the actions of some creationists are admittedly intended to influence the education of children—harmfully, in the view of a number of people. But there can be dispute over what actions should be taken about education without ruling as illegitimate the *beliefs* of one of the parties.) Quite in general, it is not the case that, because science has changed its mind in the past, therefore it might change its mind again *in any direction and by any amount.* But it is also not the case that scientific opinion is necessarily always better than popular belief.

The popular view, then, that science deals in facts, is wrong. Indeed, scientists themselves show their admiration for other characteristics of science than its facticity. Scientists are acclaimed for brilliance, creativity, originality, much more than for the ultimate in precision; publications are read to the degree that they are "interesting," not to the degree that they seem to be the last word on a matter. Of course, admiration comes for brilliance, originality, and the like only when the work also seems to be not wrong; but since all science can expect to be superseded, that is a matter of degree only.

Society as a whole does prize science because of the truth that it is thought to bring, and with it the power to control human environment and circumstances. It is tempting to use that popular view as a basis for getting support for scientific activity, and as a basis for convincing people that one is right about any number of things; but that is essentially dishonest, and I do believe that in the long run honesty is actually the best policy. The honest truth is that science does not deal in absolute facts. It is very reliable—but with exceptions, and we cannot always be sure where or when we shall encounter the exceptional. Science is, sure enough, a better guide than folklore or mysticism—in most circumstances, that is to say, and assuming that what one wants to know has to do in some way with material things. It is undoubtedly perverse to express doubts indiscriminately or fundamentally about the corpus of contemporary scientific knowledge. But it is also unjustified to claim that any given scientific notion is unquestionably right, in all its current connotations, or even that it is necessarily right when it contradicts some popular piece of folklore. Each issue needs to be looked at in its own right; and a human right to believe improbable things ought to be respected.

Scientific Knowledge as a Map

If science does not deal in facts, how reliable then is scientific knowledge? How can we be sure that it reflects something of a real world? Could it not have been arbitrarily constructed by generations of scientists, so that alternative sciences could be just as valid?

The trouble with thinking in terms of facts and theories is that we regard them as distinct things whereas they are actually connected inextricably. Science does contain enormously reliable knowledge, but it is not of the sort "The thing A exists" (a fact) or of the sort "B causes C to happen" (a theory); rather, scientific knowledge is of the sort "When one (anyone) does P, then Q happens" (almost all the time, under certain circumstances). Point a telescope at Jupiter and you see a large bright spot together with a variable number of smaller and less bright ones whose relative positions change (if the telescope is true enough, and if atmospheric conditions permit). *That* is the fact of the matter. It is a bit less factual that Jupiter has a certain number of moons, for "moon" has connotations about which we remain far from certain—for instance, how and when the moon(s) came to be associated with the planet.

Scientific knowledge is like the knowledge embodied in a map: "Follow this route, and you will pass a valley and then a river." That maps work convinces us that the landscape exists outside our mind or imagination; that science works convinces scientists (even if not all philosophers, let alone social scientists) that there exists a real world that is not the creature of human imagination. But scientific knowledge is no more than a guide to reality; it is not the real thing itself.

Philosophers in particular harbor legitimate doubts, for example, that there really exist things that correspond to what we call electrons; or that it is reality rather than the human mind that deals in something we can call energy. Much of the manner in which we think about the world remains doubtful and subject to change; but that does not vitiate the plain fact that we know an enormous amount, that we can take an uncountably large number of specific actions with almost entirely predictable consequences.

Like scientific knowledge, maps can be totally reliable guides even when they are entirely schematic. A commuter map of the Washington, D.C., metro system (fig. 8) is nothing like the real thing in size, relative proportions, noise, color, or anything else—except the relationship between stations: if you want to get from one station to another, this map is completely reliable and all that you need. Much of science is just as certain as that, so long as you stick within the limits of proper application of its maps.

Maps, like the sciences, are not collections of facts; nor are they entirely theoretical. They are some sort of amalgam of fact and theory that is staggeringly reliable without being guaranteed 100 percent forever reliable. The amalgam is always being modified. Places get renamed, different relations are recognized, above all finer and finer details are explored; and one can see no end in sight to further exploration.

Maps seem to contain more information than was used in drawing them, and often that additional information is sound. Most often, it turns out, knowledge *interpolated* from known features is reliable: if one can get from A to B without crossing a river, and if one can, some little way north, get from E to F without crossing a river, then most likely one can take an in-between route C-to-D, north of A-to-B and south of E-to-F, also without crossing a river. Or, the melting point and the

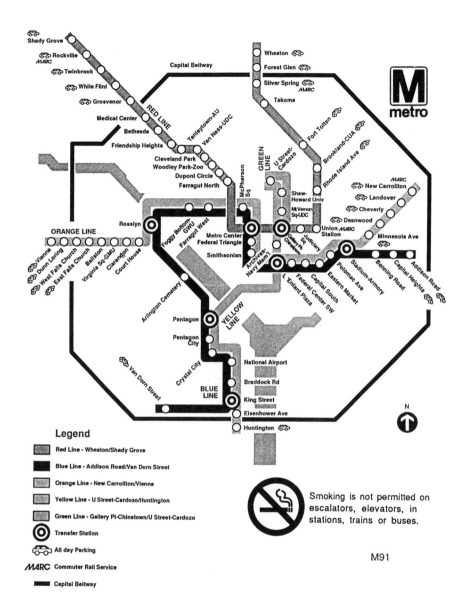

Figure 8. A commuter map of the metrorail system, Washington, D.C., reproduced with permission of the Washington Metropolitan Area Transit Authority.

boiling point of C_2H_5OH lie most likely between those of CH_3OH and C_3H_7OH.

When one *extrapolates* from the known into unexplored territory, however, great caution is in order. One cannot predict how far a mountain range stretches; one must explore it. The "revolutionary" episodes in science are of that sort: they do not really disturb the existing knowledge of "Do this, and that happens," but they reveal it to be so only within a certain range of conditions or terrain outside which the flora and fauna are strikingly different. Newtonian mechanics is just as valid now as it was 300 years ago, but we now know we cannot use it whenever speeds approach that of light (then we must use relativity theory) or whenever "things" get small enough (then we must use quantum theory).

It is not always possible to know that one is invalidly extrapolating instead of validly interpolating. Variables not previously experienced may apply: one ought not to be blamed for not foreseeing that the metro map will not work at 4 A.M. One should recognize, on the other hand, that the map is not a useful guide to a walking tour of Washington. And—though there is no good reason to bear it always in mind—one knows that accidents and strikes happen occasionally; and that, almost inconceivably yet possibly, an earthquake or a nuclear strike would vitiate the map's validity altogether.

Because scientific knowledge is a map, not the thing itself, it cannot be used with equal facility by everyone. According to John Ziman:

> In principle, every fact or theory known to science is contained in the public scientific literature . . . ; in practice, this information is only intelligible to a specialist in the relevant field.

> The public scientific archives . . . seldom give a clear response to the question "What does science know about *X*?" The primary literature may be confused and contradictory, the secondary literature equivocal, and the tertiary literature dogmatic but out of date. . . . Much . . . is only really understood by scientists themselves, as a consequence of their research experience. Thus the notion of a "scientific consensus" on a particular point . . . is ill-defined, and should be considered an ideal . . . rather than an achieved reality.

The common misconception that science deals in facts leads the media and many others—notably, pseudoscientists—to cite "facts" from the scientific literature in thoroughly misleading fashion. Immanuel Velikovsky, for example, ripped innumerable items out of scientific context to produce what was gibberish and yet seemed to many people scientific. Currently, many dietary recommendations are similarly unsound, claiming for the whole population validity that, at best, applies to particular and unrepresentative groups. Scientific knowledge, be it gained methodically or by filtering, is map-like rather than fact-like.

Successful Prediction Proves a Theory Right

When someone's prediction proves to be right, we tend to be impressed. But that a prediction has turned out right is, logically speaking, in itself neither interesting nor significant. *How often* are this person's predictions correct? Can he predict the weather or only what his wife will say when he gets home late? If only one prediction in a hundred is right, then we are unlikely to look to that prophet for advice; if he can predict his wife's behavior in situations that happen all the time, we do not even regard him as a prophet, whereas we would if he could predict the weather with some accuracy.

So too with scientific theories: that some predictions have been successful is in itself no guarantee that the theory on which they were based is generally reliable or that it is comprehensive—let alone that it is true. Almost any scientific theory, after all, was at some time derived from (or justified on the basis of) evidence. So long as that evidence remains uncontradicted, it continues to provide support for that theory; and, moreover, one could very likely continue to get more evidence of the same sort, in further "confirmation" of the theory—in other words, one could continue to make successful predictions. Thus the advent of modern chemistry is often reckoned from the rejection of the incorrect theory of phlogiston. According to that theory, when a substance burns it releases phlogiston, the substance whose chief property is heat. An enormous number of chemical reactions were classified ("understood") under this theory, and the theory was able to predict a great number of other reactions—as it still can. Nevertheless, Antoine Lavoisier was able to convince the community of chemists that burning (in air) actually signifies combination with oxygen, not release of phlogiston.

And so it is whenever a new theory is adopted: the old one does not lose its success, it is just seen to be not as successful or fruitful or compatible with other bits of science, or usable only under limited circumstances, by comparison with the newer interpretive scheme. For another instance, significant in the advent of modern science was rejection of the idea that the Sun travels around the Earth: Copernicanism, heliocentricity, the idea that the Earth travels around the Sun, became accepted instead. That was not because the new theory could make better predictions, however, but because the theory had other apparent advantages. Further examples have been given by Stephen Brush in *Science* and *Eos,* showing in particular that theories might not be accepted despite a remarkable record of successful prediction; in other words, if the theory does not fit with the prevalent conventional scientific wisdom, it may be ignored. Thus the astrophysicist and Nobelist Hannes Alfvén has considerable publications and successful predictions to his credit, yet his theoretical suggestions have not been accepted by most others in his specialty.

Such resistance in the face of predictive success is not uncommon in mainstream areas of science, though it is hardly routine—making predictions and testing them is common practice in science, and successful predictions are usually taken to mean that the theory can continue to be used. But successful predictions *as such,* for example, predictions made outside the mainstream of science, particularly in what one can call fringe science or pseudoscience, are not given any weight at all by the scientific community—generally because the ideas on which the predictions are based make no connection with textbook knowledge. Thus Velikovsky and his supporters made much of the fact that he had predicted radio signals from Jupiter, the extent of the Earth's magnetic field, and the high temperature of Venus; but the scientific community was quite unimpressed because Velikovsky's premises contradicted *all* accepted ideas about the stability of the solar system over the past thousands of years, and because he never set his theory down explicitly enough that others could judge whether the predictions really followed logically and inevitably from the theory.

Again, the scientific community pays no attention at all to claims that psychic predictions have been successful: because those claims fail to mention how often a given psychic's predictions have *not* been right; and because no explanation is given of how the predictions are arrived at; and because different psychics rarely come up with the same specific prediction. Theories must be described in detail before they will even be looked at by the scientific community, which also expects that predictions from a theory be predictions that any competent individual would arrive at starting from that theory—in other words, that a theory should not lead different people to contradictory predictions for any given event or phenomenon.

Occasionally it is suggested that *surprising* predictions that turn out to be correct do tend to prove a theory right, even if ordinary or routine predictions do not. But Velikovsky's predictions were certainly surprising, and those of psychics sometimes are too, and still they carry no weight. To influence the scientific community, a thing must fit in some way with the conventional wisdom and it must seem useful or interesting. So if a theory has won wide attention already and has not been judged inherently implausible, and if it then predicts something that other theories have not (that is, something surprising), and if that turns out correct, then many scientists will indeed be impressed and adopt the theory at least provisionally, because they will hope to make similarly good use of it. But if a theory seems to contradict significant amounts of textbook knowledge, no amount of successful predictions will lead to its adoption so long as other theories provide reasonable explanations.

A theory's ability to make predictions, then, is certainly regarded as a good thing. The more surprising the predictions, the more we tend to be impressed. But the flat statement that successful predictions prove a theory right is quite wrong and misleading.

Science Is (or Should Be) Open-minded

Scientists and true believers in science are fond of pointing out that science is always open to new ideas, new methods, new knowledge: how otherwise could one explain the progress of science, which has seen the continual acquisition of new knowledge by deployment of continually new methods and the adoption of continually new theories?

This is the same fallacy as that which led to the myth of the scientific method. Hindsight gives a misleading perspective. There may not be a simple cause for what has happened in science in the past few centuries. One might equally explain the course of biological evolution through openness to change, just because as we look back we see that there has been so much change. In point of fact, we believe that biological change is anomalous rather than normal. Reproduction proceeds by duplication of existing genes, with considerable safeguards against errors in duplication, so that organisms on the whole breed true; the challenge is much more to explain variation and novelty than faithful heredity and stasis. The particular course that evolution has taken is a *by*-product of mutation, competition, environmental change, natural selection, genetic drift, and the like, *not* the result of any direct natural tendency for biological entities to change. Similarly, that science has adopted new ideas in no way demonstrates that the adoption of new ideas is somehow fundamentally natural to science. Indeed, to assert that science is open to new things is to fly in the face of the evidence; and it even contradicts other common beliefs about science.

To make sense of the tension between innovation and conservatism in science, more helpful than the banal distinction between what is known and what is not known is the discrimination of three categories: the known, the known unknown, and the unknown unknown.

The known comprises not what we know but what we believe, what we *think* we know; it is the conventional wisdom, the governing or prevailing paradigm. That paradigm not only embraces what we think we know, it equally determines what we believe to remain unknown. For example: the periodic law recognizes that, when the chemical elements are arranged by increasing weight of their atoms, those elements with similar chemical properties recur periodically (the third, eleventh, and nineteenth elements react very similarly; so do the fourth, twelfth, and twentieth; and so on). But when this was first recognized, there was a clear implication that certain elements with certain properties remained to be discovered. Almost immediately, research into this area of the known unknown, guided by the known, led to the finding of some of these previously unknown elements.

The known unknown, then, contains known *sorts* of things and phenomena, though we are ignorant about many of their details. Having become clear that DNA transmits information from one biological generation to the next, it became clear at the same time that the next steps to be taken—the next pieces to be added to the jigsaw puzzle—were to find out how DNA turns that information into instructions that certain substances shall be made at certain times, and that cells shall specialize at certain times and in certain ways.

The unknown unknown comprises what we do not even suspect. Indeed, were it not for history, we would not even believe that the unknown unknown exists. We learn the prevailing paradigm, and thereby not only what is known but also what is believed to remain not understood. The conventional wisdom is blind to its own inadequacy, to the fact that, sooner or later, it will be found to be wrong, in one way or another. Each generation of scientists has believed that it understands—even if only in broad outline—the chief principles that govern natural phenomena. Following Newton, many generations were convinced that all physical phenomena are just matters of particles and forces. New types of forces were discovered, to be sure, but that did not shake the fundamental conviction that physics could proceed along foreseeable lines of research to discover everything that could ever become human knowledge. By about 1870, scientists felt quite secure about the main principles with which all phenomena could be explained. But within a few decades, entirely out of the blue, came radioactivity, the discovery that some atoms self-destruct; and then the necessity to describe radiation sometimes as particles, as packets of energy, rather than as waves; and then relativity, non-Euclidean space, the uncertainty principle: a succession of total surprises, as revolutionary as an intellectual revolution ever could be.

Nowadays, again or still, even the most forward-peering scientists believe that all the main principles have been recognized and that no major surprises await us. Fred Hoyle—in many respects an iconoclast and enfant terrible—declared in 1960 that "by and large . . . our present picture will turn out to bear an appreciable resemblance to the cosmologies of the future." And Carl Sagan, in many respects incisive and critical, wrote in 1978, "This . . . is written just before—at most . . . a few years or a few decades before—the answers to many of these vexing and awesome questions on origins and fates are pried loose from the cosmos. . . . there is only one generation privileged to live through that unique transitional moment: that generation is ours." Thus, human beings, including scientists, do not function under continual awareness of humanity's fundamental ignorance; rather, they live under perpetual illusion of fundamental understanding.

Thomas Kuhn's "normal" science is governed by the conventional wisdom and is what happens most of the time in science: it is planned as research into the known unknown and any surprises are expected to be manageably minor ones, matters of detail within the paradigm and not new principles. Revolutionary science comes about when something out of the unknown unknown thrusts itself unexpectedly upon us: that particles are also wave-like, and that waves are also particle-like; or that parity is not conserved, that there is something fundamentally asymmetrical about our universe.

So one can comprehend that innovators in science *routinely* encounter resistance if their ideas are sufficiently original; almost invariably, if those ideas contradict significant parts of the conventional wisdom; the more strongly, if novel methods are also involved; and, of course, more emphatically if the innovator is relatively unknown or an outsider to the relevant specialty. The list of discoveries resisted (only to be eventually accepted) is long indeed, and the names read nowadays like an honor roll: Ampère, Arrhenius, Copernicus, Einstein, Heaviside, Helmholtz, Lister, Mendel, Ohm, Pasteur, Karl Pearson, Planck.

Lesser lights than these also find that their most original ideas or results are the least palatable to their peers. It is simply the case that human beings, be they scientists or something else, do not take kindly to having their beliefs contradicted. Admittedly, discovery in science is given lip service as desirable, and scientists strain to achieve it: but what they aim for is discovery *within the prevailing paradigm only.* The genuinely novel arises unbidden and unforeseen out of the realm of the unknown unknown. Often it comes at the hands of a maverick or an outsider who is not hampered by the disciplinary blinders of the specialists, and inevitably it encounters incredulity, to some degree or other, at least at first. Science is open to new things only so long as they are *not too new.*

This resistance to novelty is not any flaw in science that arises from the fallibility of the human beings who are doing science. Far from it. This resistance is actually the foundation of the trustworthy strength of science. The conservatism of the scientific community ensures that science itself is conservative and conserved, that new notions must prove themselves quite compellingly, with overwhelming evidence, before they win the day. It is from this that the reliability of scientific knowledge stems. Before novelty can be published, it has to convince referees and editors that it is not obviously wrong. Before novelty is listened to in the scientific community, it must seem well founded and plausible. By the time an erstwhile novelty becomes textbook science, it has stood the tests of natural conservatism and time and it is much the more robust for it (of course, many novelties do not survive the tests and just fade away). If science were wide open to new things, then science would be quite unreliable. Frontier science, which is comparatively open, is also notably unreliable.

But even at the frontier it would not be accurate to say without qualification that science is inherently open. In frontier science one has competing scientists whose minds are open to varying degrees to various things, a state that could be described with equal accuracy by saying that these scientists' minds are *closed* to varying degrees to various other things. Science progresses not because scientists as a whole are passionately open-minded but because different scientists are passionately closed-minded about different things.

What about prizes in science? Are they not given for novelty? Does not the greatest novelty receive the greatest prize? How can that happen if science is basically conservative? Again, because it is only the novelty that has stood the test of conservative resistance that gets the prizes. Novelty alone is not enough: not *any* novelty gets a prize, not *any* novelty is applauded, as would be the case if *novelty in itself* were prized in science. So too with human societies as a whole, of course: revolutions and revolutionaries are honored, but only in the abstract or in the past. Actual contemporary revolutionaries are resisted, be it in Latin America or Africa, in the Western democracies or in the Eastern people's democracies; while at the same time there is ritual honoring of Bolívar and Nasser, of the American Revolution and the Bolshevik one. Revolution, in science as in political affairs, gets prizes only after it happens to have won the day. Novelty receives a prize only after it has become part of the conventional wisdom.

Resisting revolutions makes good sense, because the results of attempted revolutions are always unpredictable and often unwelcome. In science, history demonstrates that maverick notions have almost always been unsound. But history demonstrates that only if one looks closely and deeply, because unsuccessful *un*orthodoxies tend to disappear, leaving little or no trace. Superficially, history can mislead—and has misled—because it keeps recalling the *rare* troublemaker who turned out to be right (Galileo, Semmelweiss, etc.); it forgets the scores or hundreds of would-be iconoclasts who were also quite sure they were right but were not, and who were with full justification quashed by the wet blankets of orthodoxy.

To assert that science is open-minded, or that it should be, not only contradicts the evidence but also clashes directly, and not too subtly, with other common beliefs about science, in particular that scientific knowledge carries a notable degree of certainty. Everything that one learns has not only a positive side—that a certain thing exists, that a certain phenomenon takes place—but also a negative one—that certain other things therefore cannot exist, that certain other phenomena therefore cannot occur. This, of course, is the source of resistance to novelty. Having *explicitly* learned certain things, scientists and science have at the same time learned *implicitly* that other things are not so. We could then be truly open-minded only about things that we do not yet know about *at all,* or things about which we know so little that we cannot even judge their plausibility. That is an empty sort of open-mindedness, good in lighthearted bull sessions, perhaps, but irrelevant to organized knowledge seeking. The only truly open mind would have to be at the same time a truly empty one. Thus, even the assertion that science *should be* open-minded makes little sense. As G. K. Chesterton pointed out, the only good reason for an open mind, as for an open mouth, is that it be closed again on something solid.

Much about the actual everyday practice of science can be understood on the basis that science is a conservative, even hidebound enterprise; little, if anything, can be understood about the actual practice of science by regarding it as inherently open to new things. Indeed, genuine absurdity results if the demand that science be open is taken to logical conclusions. Thus, during the Velikovsky controversy some social scientists maintained that, because science should be and must be open, therefore there exists a right to be published—no matter how heretical the material, no matter the degree to which it seems incompetent to the rest of the scientific community. And even further, the claim was made that there exists a *right to be read,* because new ideas obviously cannot be judged if people don't know about them. The logic was impeccable, the conclusions absurd; ergo, the assumptions were wrong.

Scientists do not consciously aim for *fundamental* novelty. They seek to extend knowledge from its existing base, to bring more of the unexplored territory under the rule of the governing paradigm. They look for problems that are soluble, for things that will attract attention, for fields sufficiently accepted that money is available to support research.

Scientists Should . . .

A whole host of little fables about science begins with the phrase "Scientists should . . .": they should make their work publicly available; they should publish all their data; they should give due credit to others; they should be moved by the desire for knowledge, rather than for fame or money; and so on. All these fables contain an element of insight, but all of them are also misleading if taken literally, for they embody mistaken views about human beings and about science.

. . . Publish All Their Data

As the analogy of the jigsaw puzzle makes plain, science is a communal process that works best when all the players know all the moves the other players make. Naturally, then, all the players should make their moves publicly. "Secret science" is a contradiction in terms: it could go only a little way before coming to a halt because of the absence of pieces held by other players who are not part of the secret group.

But not all players are equally competent; some of them persist in trying to force pieces of the puzzle into places where they don't really belong. It would be better if such moves were not widely known, for they could only mislead; it is better for players to be sure that their moves are correct before they trumpet them to the world. For a while they should keep quiet, trying the schemes out only on their close colleagues, until they are reasonably certain that their schemes really work and are not just wishful thinking. So, yes, scientists should indeed make their work publicly available—but that does not mean they should make available every item of information or piece of putative data as they obtain it. Scientists must always judge whether it would be potentially misleading to make some of their activities publicly known, and they need to keep those activities private until they believe they have something sound to contribute.

Moreover, when scientists do make things public, they must not then publish all the data they have acquired along the way. There will have been false starts, experiments that turned out to be flawed, inexplicable deviations every now and again that pointed to something going wrong without it being clear exactly what. Scientists should publish everything that they believe could be of value to others, but they should not publish data that (they believe) could mislead others. In other words, they should publish all the data they believe to be sound.

Of course, what is believed to be sound is plainly a matter of judgment. One's view of the soundness of data is influenced by the beliefs one holds about the *meaning* of the data. One scientist, adhering to a given theory, might well judge unsound the data that another, holding a different view, regards as sound. Thus in the early years of the century there was dispute over the nature of electric charge: Robert Millikan believed it to be particulate; Felix Ehrenhaft was not so sure. Ehrenhaft published data that indicated the existence of fractional charges; Millikan published only data that showed integral charges. In recent years, reexamination of Millikan's original notebooks has revealed that many of his measurements could have been taken to indicate nonintegral charge; but he had found—or suspected—something wrong in those experimental runs. On the other hand, runs in which integral charge was indicated were annotated with such comments as, "Beauty, publish this surely, beautiful!"

Some contemporary pundits have suggested that this selective publication by Millikan constituted misconduct, if not outright fraud. Yet if Millikan had set out to mislead deliberately, he would hardly have left notebooks available that so clearly document his actions. Surely Millikan believed that he was doing the right thing by rejecting results that were flawed and publishing only those that seemed sound.

Admittedly, Millikan left unavailable to others data that might have supported an alternative theory. But that is simply an unavoidable dilemma. To ask that every scientist publish every piece of data is to invite a flood of unsound, uninteresting, misleading garbage. Research papers are and should be judged by the new understanding they contribute, not by how many undigested measurements they contain. Refereeing would become impossible if one took literally the shibboleth that "all data should be published"; indeed, if that were the practice, then refereeing would have little purpose.

. . . Give Credit to Those Whose Work They Build Upon

Surely the injunction "Scientists should give due credit to others" is one that needs no qualification; it is just a simple matter of ethics. But no, even a little thought about actual practice makes plain that judgment must be exercised in citing the work of others. Consider once again DNA, Chargaff, and Watson and Crick: how much credit—and what part of the Nobel Prize—was due Chargaff for his measurements (the raw data) that undoubtedly gave Watson and Crick evidence to support their structural model (the crucial *interpretation* of the *meaning* of the data)?

In 1962, a letter in *Science* suggested that scientists ought to credit the priority of Velikovsky for the statements that Venus is hot, that Jupiter is a source of radio signals, and that the Earth is surrounded by a substantial magnetic field. Yet Velikovsky had made those claims only in popular books, not in the scientific literature. Further, the claims were not *based* on anything in the scientific literature; rather, they were based solely on Velikovsky's idiosyncratic interpretations of literary and historical texts. *Why* should Velikovsky have been given credit? Those who discovered the Van Allen radiation belts, the radio emissions from Jupiter, and the temperature of Venus had been entirely uninfluenced by Velikovsky's predictions; indeed, they had been quite unaware of them. And since these claims had never been made within the scientific literature, one could not even say that they *should* have been aware of them.

Should Ernest Rutherford, generally regarded as the first person to observe the transformation of one chemical element into another, have paid his respects to the long tradition of alchemy in which methods were sought to transform base metals into gold? After all, to alchemy belonged the important concept that elements *could* be so transformed, a concept that had subsequently been rejected by science for a couple of centuries. When Rutherford was seeking explanations for his results, might he not have been subconsciously led to think about possible transformation because, through the prior existence of alchemy, the concept was already familiar to him?

Giving credit means avoiding plagiarism. It means acknowledging what has meaningfully influenced one's own work. It also means referring the reader to other work that one regards as sound. And all that inevitably involves making judgments. So the folklore and the history of science are replete with disputes about whether or not adequate credit has been given, and it is by no means only that obscure scientists may fail to be cited through sheer oversight. As mentioned earlier, Stephen Brush points out that the Nobelist Alfvén is often not cited when it would seem that he ought to be.

As historians learn in more painstaking detail about the development of science, it becomes ever more clear that discontinuities in the history of ideas are rare, if indeed they exist at all. What seems at first a brilliantly creative novelty turns out to have precursors, albeit perhaps only by analogy or metaphor, but precursors nevertheless: *nihil ex nihilo* may well be a fundamental law of human thought. The traditional emphasis on discovery and originality, however, serves to mask that continuity. The existence of prizes in science also helps imprint the mistaken notion that discrete discoveries—distinct breaks with past practice or thought, in other words—are possible. On the other hand, the argument that so often surrounds prizes, about others who ought also to have been honored, illustrates how difficult it is even in superficial practice to maintain that notable advances stand alone.

Scientists Cannot But Make Judgments

Nothing of note in the practice of science is entirely mechanical, contrary to the popular formulation of the scientific method that makes it seem mechanical; judgments must always be made. As in all human endeavors, the soundness of judgment can be assessed reliably only in retrospect, if then. Scientists who turn out to have been wrong might be victims of bad luck, or of their own poor judgment, or of simple incompetence—or they might have been frauds. Before one decides that a person has been guilty of fraud or misconduct or breaches of ethics, the possibility of incompetence or poor judgment surely ought to be conclusively eliminated. In the contemporary fuss about fraud in science, that is seemingly not understood, and accusations of misconduct are made—and publicly at that—solely because some piece of work turns out to have been imperfect.

Here again the myth of the scientific method is a contributing factor. Proper scientific method, according to the myth, means setting up well-defined hypotheses, designing appropriate tests, carrying those tests out carefully, assessing the results objectively, and publishing the conclusions. When it is put in those terms, why should anyone not be able to follow proper method? The prescription is transparently straightforward. It seems natural, then, to see it as avoidable wrongdoing whenever individuals fail to live up to the prescription.

The rub, of course, is that this prescription of method makes no literal, practical sense. It is not worth testing hypotheses unless they have to do with scientifically interesting things, with matters not clearly understood yet sufficiently in the realm of the known unknown that they may be accessible to current technique. One's career depends on a reputation for originality: because the opportunity to do science depends on getting jobs and grants and facilities, one has to impress people with one's capabilities, to show that one can do things others cannot. One must make judgments all the time: about what research to do,

about when support for one's views is strong enough to warrant publication, about which details to publish and which to withhold—either
because they might mislead or because they might enable others to
anticipate one's next moves and thereby preempt one's own claim to
priority and credit. Popular descriptions of the scientific method portray
a formula that anyone ought to be able to follow in quite mechanical
fashion; yet nothing of any importance in science is mechanical. Rather
than chide scientists for not following the method, we should emphasize
that scientists, like all other professional people, should strive to the
utmost to behave ethically, for without that science will not work properly.

Science Is Self-correcting

I have just illustrated how the myth of the method can cause unfounded accusations of improper behavior to be directed at some scientists. Conversely, at the hands of scientists, the myth of the method
can produce unfounded confidence that the scientific community need
do nothing to keep itself honest: nature and the scientific method are
supposed to do it automatically.

However, if science is properly understood as one particular human
activity, then it becomes clearer what scientists "should" really do: they
should behave ethically. So far as possible, in their professional actions
they should avoid conflicts of interest, wishful thinking, the temptation
to take shortcuts. They should judge their colleagues and peers on
professional grounds, not on the grounds of friendship, nationality,
gender, or other irrelevancies. And so on. As with lawyers, doctors,
politicians, and other human beings, scientists should strive to follow
proper standards of professional conduct.

In other professions, codes of conduct have long been explicit; in
medicine, for example, the Hippocratic ideals date at least two millennia
into the past. Science, by contrast, is modern. Only in the seventeenth
century did science become a thing more or less recognizably like what
it is today. Only in the nineteenth century did science become a profession, an activity by which one could earn a living. The very term "scientist" only came into use toward the middle of the last century. In
more recent decades, science has come to involve large numbers of
people, not the relatively small number of relatively elite individuals
who created the age of science a century ago. Only in recent decades,
also, have the actions of scientists come to have so important an effect
on the wider society that manifold details of scientific activity have come

under public scrutiny. And so it is only nowadays that one sees so clearly the need for scientists to have an explicit code of professional conduct.

Having come under public scrutiny, science is no longer taken automatically to be a good thing, something that society should support as a matter of course. So scientists must—if they wish to be supported—convince others that what they are doing is worth the support, and that they are making good use of it. That means making assurances that the profession is able and willing to influence its members to abide by the publicly stated ideals of the profession.

Just as has long been the case with medicine, society expects that the profession of science itself will take action against incompetent conduct and against unethical conduct. Having long made the claim that science must be left to govern and police itself, science must now convince its audience that it is doing those things effectively. Just as in medicine, if science itself is not willing to distinguish incompetence and fraud from humanly inevitable poor judgment, then the wider society will try to make that judgment—though the wider society is less competent to distinguish humanly poor judgment from avoidably poor practice, so that unjust consequences for individuals and deleterious consequences for the profession and for society itself would be likely to follow.

It seems to me not farfetched to compare the current state of science (and more generally that of academe) to the situation of the Church at the time of the Reformation, which has been described in the following way by De Lamar Jensen:

> Until the middle years . . . the actual number of clergy [read *scientists*] increased, but then a decline set in. Even before the outbreak of the . . . revolt, their prestige and influence were already waning. Whether justified or not, the general population's growing disrespect for the clergy [read *scientists*], especially the monks [read *researcher-scholars*], tended to weaken some of the bonds of the Christian [read *scientific*] community and make the church [read *scientific institutions*] as a whole more vulnerable to criticism and attack. It had not been above criticism in earlier ages, but now it was becoming the practice rather than the exception to blame the institution as a whole, along with individual members of it, for infractions . . . of law and . . . ethics. As . . . abuses increased, the recognition and condemnation of those abuses mounted proportionally. To compensate for their declining prestige, many clergymen [read *scientists*] became even more avaricious [asking for ever lower teaching loads, higher salaries, freedom to consult and to found business enterprises; ignoring

conflicts of interest], and the growing chasm between the priest-hood [read *scientists*] and the laity, and between the higher and lower clergy [read *administrators and practicing scientists*], widened.

The Reformation was no revolution against religion itself; rather, it was spurred by disenchantment with the people who officially professed it but in practice behaved badly. Just so is modern society disenchanted not with science itself but with the practices of some of those whose profession it is.

This disenchantment is now fed by an increasingly large corpus of books in which actual scientists and their activities are described in ways that cater to human (in the sense of prurient) interest, with emphasis more on competition, dirty dealing, ambition, and the like than on the advances in knowledge achieved. This modern genre of scientific docu-novels, where scientists are portrayed as subject to all the passions *except* dedication to the search for knowledge, may be said to have begun with Watson's autobiographical account of his race to be first with the struc-ture of DNA. As Horace Freeland Judson put it, "Watson has given his most spontaneous acts the color of calculation"; and Max Perutz, who was there, has also pointed to the misleading tenor of Watson's memoir in passing over the intensive reading and thinking that he and Crick engaged in. Accusations of misconduct by scientists have also been fea-tured with weekly regularity in newspapers and magazines. Several work-shops have been arranged by the American Association for the Advancement of Science to discuss these issues, and congressional hear-ings have been conducted.

Just as in the Reformation, society is not now in the mood to distin-guish between the human failings of individuals and the activities of the institution. Priests and scientists, of course, are just poor sinners like everyone else; but that recognition does not excuse the institution from acting against sin by punishing sinners. To understand must not be to excuse, or the wider society will rightly conclude that the institution is less concerned with its ideals and responsibilities than with the short-term welfare of its present members. Just as the Church could not count on God alone to restrain priests from sinning, no matter how consci-entiously they might pray for it, so the scientific community cannot count on the scientific method to nip in the bud any tendency to stray on the part of individual scientists.

A way through the woods, and perhaps the only possible resolution, requires that the myth of the scientific method be exposed and under-stood, by scientists themselves as much as by outsiders. Once science is seen as an institutionalized human activity, then the need for an explicit code of professional ethics becomes plain and replaces the unworkably abstract "shoulds" implied by the method.

But scientists themselves believe what philosophers and others have long told them: that the success of science is owing to the application of the scientific method. Thus, successful scientists are led to delude themselves that their own success proves that they have faithfully followed the method. Criticism from outsiders is then seen as *uninformed* criticism and is readily ignored. Scientists themselves believe science to be self-correcting, because the philosophers and the historians and the scientific method have told them so. Scientists have seen no need for a code of professional conduct because they have believed that the constraints of reality, exercised through the method, have been stronger and more automatic in application than any human code could be.

Now it is the case, of course, that nature constrains science, but only over the long run and not quite automatically. Society, however, is always concerned with the present and the immediate future, so the scientific community must demonstrate ethical and effective behavior *in the short run*, where the constraint of nature can do relatively little to weed out error or fraud. Only if the profession convinces society of its effective ethical concerns will society allow the profession to continue to govern itself.

Science's claim that it cannot effectively be controlled from outside itself is of course true—but only in the sense that good science cannot be produced on demand, that the answers society wants may not be achievable, that knowledge provided by science may be unpalatable. Society can certainly control the funds it provides to science, and it can control much about the life that scientists lead; and if that stultifies science, well, that would be felt much more strongly and deeply by the scientific community than by society at large. Human societies existed for a long time without benefit of science; and many contemporary human societies continue to do so. It might be a rather poor, rude, uncivilized existence, but science cannot count on the fear of that consequence in its striving for public credibility and acclaim. The times call for publicly stated ethical guidelines to which the profession holds its members.

Great Scientists Can Speak for Science

Science is popularly seen as a thing rather than as a collection of loosely related sciences and sorts of science; what supposedly makes it a thing, of course, is the scientific method. It follows that scientists who have won prizes or who have been singled out for other honors must be the most adept at using the method, and therefore they will be the appropriate people to speak to the wider society about anything and everything having to do with science.

Of course, all common sense and human experience, as well as proper understanding of scientific activity, demonstrate just how false that is.

In the first place, one's expertise in any given practice does not transfer readily, let alone automatically, to another field. If only it were widely enough understood that science actually comprises a myriad of overlapping, but still significantly distinct, specialties, then the public and the media might be clear that one who has a Nobel Prize in physics is not inevitably thereby worth listening to on points of chemistry, biology, sociology, or politics. Indeed, Nobelists can be just as ignorant as anyone else, outside their sometimes *very* narrow expertise. Thus a Nobelist in medicine has given popular talks about the origin of life in which he makes elementary errors about chemistry (that the ability to form double bonds is restricted to elements in the first row of eight in the periodic table), and one in physics talked simplistic nonsense about the racial heritability of intelligence. As already pointed out, since scientists themselves share belief in the myth, accomplished scientists sometimes fall into the delusion that they have mastery of a method that enables them to be wise in all fields.

Perusal of the things for which Nobel Prizes are awarded reveals that many of the accomplishments seem much less striking a decade or two later, but chiefly that most of them have to do with what, again by hindsight, one sees as rather minor detail. Successful scientists are in the right place at the right time. They do have talents that are sometimes quite extraordinary, in insight and judgment and ingenuity, and almost invariably they are extraordinarily dedicated to what they do. But those talents are evident in very particular practice, almost always in a very restricted portion of one of the many scientific specialties, and those talents may be irrelevant to other practices. That one has been hugely successful in chemical synthesis, for example, does not mean that one has anything of interest to say about other matters. Indeed, the more intelligent Nobelists are fully aware of this, and of the unwarranted demands the media tend to make on them, and they refuse to use their fame as a stepping-stone to generalized guruhood.

Great discoveries in science are not necessarily made by people whom one would describe as "great" in some overall sense. There is the element of luck—having had just the right training, just the right and unusual combination of experience, and so on. There is the Zeitgeist: the time may be ripe for something, the Jigsaw Puzzle is just asking for this next move—and so discoveries are often made almost simultaneously by several people independently (and acrimony over priority is common). There is the thread of continuity that we can trace, at any rate by hindsight, through which precursors of some sort can always be found. Scientific creativity does not bear the creator's stamp in the way that artistic or technological creations do; to repeat Erwin Chargaff's aphorism, "With very few exceptions, it is not the men that make science; it is science that makes the men." Science is very much a communal activity, and individual scientists become greatly honored for a variety of reasons, but most commonly for very specific contributions rather than for any display of wide erudition or deep wisdom.

On the whole, the most technically brilliant scientists are not usually those who have the best perspective over larger matters, perhaps in part because science demands such intensity of concentration, such single-mindedness, that it is difficult to have time to think about other things. If one feels the need to find people who will speak globally on behalf of science and its role in society, one would do better to turn to people well respected for their institutional rather than for their technical contributions. Robert Oppenheimer, for example, was enormously admired for his leadership of the atomic bomb project, but he was not in the ranks of Nobel-class physicists. Since the most remarkable advances in science have a large element of serendipity, a single major discovery is all that most eminent scientists manage; one like Linus Pauling, who is rightly credited with several great contributions, is very rare indeed. On the other hand, scientists who have made their mark as administrators or organizers have had to demonstrate ability and achievement fairly steadily over appreciable periods of time, and have had to show that they can balance technical imperatives with human capacities and individual peculiarities with institutional demands—in other words, that they have a somewhat wider perspective.

Almost always, what the media or the public want from science has to do with applications, or with technology, or with sociopolitical implications rather than with the technical science itself. Then, it is not just that the most technically brilliant scientists may not be the best authority, but even that scientists as a class may not be the best authority. As should already be evident, I would suggest that historians of science— or, even better, professional students of science, technology, and society—would be the most likely to evince authentic insight into these broader issues. The prevailing tradition, however, is that if one wants to know anything about a matter pertaining in any way to science, one should ask a scientist.

5

Imperfections of the Filter

In science, new claims are constantly appraised in the light of existent knowledge, by fallible human beings organized in particular ways. Thus scientific knowledge is only an imperfect map of the actual world, imperfect because
—existent knowledge may be misleading, particularly where striking novelties are concerned;
—individual human beings cannot be entirely objective; and
—collective human institutions work imperfectly.

Objectivity versus Consensus

The knowledge filter produces consensual knowledge, which is not the same thing as objective knowledge. Nevertheless, the scientific community's views are constrained by what experience has already shown can happen and by what cannot happen: the maps against which claims of new exploration are judged do usually reflect something of the actual landscape. According to the myth of the scientific method, scientific claims are directly tested against reality; under the puzzle and filter analogy, claims are still tested against reality, it is just that the testing is somewhat indirect.

The false notion that because absolute objectivity is unattainable, therefore scientific knowledge is as much a matter of human opinion as is any other form of knowledge, no more solid and no more valid, is promulgated by various cliques. The most common technical label for these cliques is "relativist," standing for the belief that all knowledge is only relative; but there is no essential difference between this and solipsism or know-nothing-ness. Scientific knowledge, so goes this relativist line, is "constructed," not "discovered"—agreement about what is known is "negotiated" among human beings; in science, just as in other human activities, "anything goes" that people can get away with; good science and bad science cannot be distinguished because no absolute criterion is available for doing so; we may talk of "change" in science but we cannot validly talk about "progress," again because of the lack of a stable criterion.

Those relativist notions could stand only if the scientific consensus were quite unconstrained by nature, and that is patently not so. Science remains the study of nature; *rational* opinion finds science the more satisfactory the more it properly reflects what nature does. The consensus of the scientific community—the consensus of *rational* opinion formed as widely as possible, as John Ziman puts it—is exceedingly sensitive to the test of nature. In Richard Burian's happily chosen phrase, scientists make liberal use of "reality therapy." Perpetual-motion machines are believed to be impossible because, no matter how ingeniously designed, they have never worked—period; pigs do not fly—period; there is no element of atomic number between that of hydrogen and that of helium—period; and so on and so forth. Our *explanations* for those truths may have little warrant, and the explanations we use do change from time to time. But that human *interpretation* of nature is always subject to change does not entail that human knowledge of nature's *phenomena* is always fragile: maps can be crude or flawed and yet perfectly reliable in important respects.

In the modern understanding of science that has discarded the myth of the method, nature does still constrain observation and experiment and thereby also interpretation (or theory, or scientific belief). It does so less directly, less precisely, less automatically, and less quickly than is envisaged in the classical formulations of the scientific method; nevertheless, nature cannot but remain the ultimate and entirely firm arbiter. Consider again Michael Polanyi's jigsaw puzzle analogy. Actually and ultimately, there is only one way to fit all the pieces together. But before the last piece has been put in place, there is no guarantee that all are in their correct positions; and in the earlier stages of puzzling, almost all the assembled little clusters may incorporate some mistakes, and the relationships among some of the clusters may be wildly wrong. The jigsaw puzzle is so excellent an analogy for science because it reflects the scientific faith that an external world exists, while allowing at the same time for the imperfections to which science, as a human activity, is prone. Reality therapy is so excellent an analogy for the same reason; and it, too, encompasses the necessarily communal aspect of science, for as is well known, few (if any) people can successfully practice therapy on themselves, unaided. And the analogy of therapy recognizes that reality does not determine belief all at once; that it is rather a lengthy, slow, nonautomatic process by which one makes one's beliefs approximate reality.

Human Characteristics

People do not always know what is good for them; and even when they do know, they do not always act on the knowledge. That reality therapy is often available does not lead everyone to take advantage of it, not even everyone who claims or thinks to be a scientist. Individual scientists quite often choose to believe something that is clearly false—just as all human beings often so choose, because that is how human minds work.

No one is forced to become a puzzler. No one is forced to become a scientist or to attempt to think scientifically. People can choose to play any game they wish. Some of those who choose to play may be incompetent, perverse, or otherwise a hindrance rather than a help in putting the puzzle together. They may miss opportunities, or dither incessantly, or spend their time compulsively readjusting the pieces others have added instead of adding new ones themselves. Perversely, incompetents or madmen may sometimes, for whatever reason, manage to find a fitting piece that eluded everybody else. Even the fussy ditherers sometimes turn out to stimulate progress: although they themselves rarely add anything new, they do sometimes discover that an unsuspected error exists. Some players may force nonfitting pieces without consciously cheating, just out of wishful thinking induced by too much ambition, say. And there will be some others who seek to cheat quite consciously, perhaps forcing a piece that is almost but not quite right, or trimming it a little, or even forging whole pieces. A few players will even hold that cheating is the name of the game.

In science, as in any other human activity, one encounters human vagaries. Those who review a research proposal are influenced not only by its perceived meritoriousness but also to varying degrees by their opinions about the proposer (competent or mediocre, friend or competitor or neither) and about the proposer's institution (one more readily recognizes merit in something from, say, Michigan than something from Podunk). One who is rushed for time may make only a superficial review; one who has failed to get grant support may find endless criticisms to make of others' proposals. Professors vary in the degree to which they exploit the labors of their students or the degree to which they disinterestedly foster their intellectual development. Colleagues differ in the degree to which they consider personal, by contrast to professional, attributes when recommending tenure, promotion, salary raises, or prizes. Scientists, like other people, have their dark side, and sometimes it shows.

Successful institutions appropriately harness human drives. Thus in science, as in other communal activities, honest and trustworthy individuals are, by and large, preferred over others for managerial and representative roles. Scientific institutions harness human energies to do cooperatively what individuals cannot do: for example, critically test a theory. One who has discovered or invented something novel is enraptured by the new toy and finds it much easier to discern support for it than to think of contrary possibilities that would crucially test it; like everyone, an inventor or discoverer wants to be proved right. But other people, competitors, also seeking to become preeminent in the field, want to prove that theory wrong, and if a way to do that can be found, they will find it. Thus *science* can approach skepticism and objectivity even when individual scientists are anything but objective or skeptical. (So too legal equity is served when opposing attorneys passionately, even unscrupulously argue their cases; so too, most economists believe, the best and ultimately the fairest distribution of products and rewards comes about when a plurality of selfish competitors is active in the marketplace.)

That human vagaries permit institutions to function fairly well does not of course entail that one need not or should not urge individuals to discipline themselves. The sociology of science is often said to have begun with Robert Merton's identification of norms governing the behavior of the scientific community. To be valid, Merton suggested, scientific knowledge has to be universal: there cannot be one physics in China and another in Saudi Arabia. Knowledge has to be shared—or it cannot be tested, in particular for universality. Discoveries have to be scrutinized within a context of skepticism, otherwise testing will not be sufficiently critical. Science has to be disinterested, otherwise false knowledge will be promulgated by ideologues.

Some interpretations of the Mertonian thesis hold that *individual scientists* should be skeptical, objective, disinterested. Certainly one can agree that these are ideals for which individuals might strive. Yet one might also wonder whether brilliant creativity is compatible with skepticism or objectivity; it is not obvious that science would necessarily flourish if scientists were somehow to become individual exemplars of scientific methodism. Science may be better served when some scientists generate novel ideas while others carp at everything new than if all scientists could somehow become disinterestedly skeptical.

The question is, what realistic standards of conduct should scientists adhere to? To be sure, even inveterate speculators might usefully strive to cull their speculations before pushing them; and moderate, by contrast with extreme, passion in the pushing could be an improvement. But there are limits beyond which one cannot expect disinterested, skeptical objectivity from people who are passionately curious to find out how the world works. Not even years of psychotherapy, after all, enable most of us to learn about most of our motivations, let alone how to govern them, let alone acquire the will to do so. The dispassionately psychotherapeutic and the ardently single-minded scientific viewpoints may not even be compatible. Thus a pair of psychoanalysts (whom I knew only briefly and socially) ventured that no human being could possibly be passionately interested in astronomy, say: the passion must somehow be perversely misdirected from a more appropriate object.

Though one cannot demand dispassionate curiosity, one can demand from scientists, as from others, honesty and integrity. We have already seen that translating such notions into scientific practice is not straightforward, however: thus honesty could superficially be interpreted to mean "sharing all one's work," whereas it would actually be misleading to publish everything one has done without sifting out suspected errors. Quite clear, however, is that scientists—like other people—find it easier to be honest when they are not tempted to be dishonest; they are more readily honest if being honest is clearly in their own best interest.

Corollary to the myth of scientific method is the classical stereotype of the scientist as solely interested in satisfying curiosity, in discovering truth; and classically, this may have been largely the case when science was pursued as a hobby or avocation. But the organization of science that began in the seventeenth century led scientists *as scientists* to be also organizers and managers, with consequently inevitable conflicts of interest. Now that scientists earn their livings as scientists, they experience an inevitable conflict of interest between seeking truth and making a career. How much time to spend writing potboilers to pad one's vita, as opposed to working on the most challenging problems? How deferential to be to some misguided program manager who will give you a grant only if you slant your work in a direction you believe not really worth pursuing? And so forth. Conflict of interest is inherent in being a scientist, as it is in being a doctor or a lawyer, say: one's income depends on the extent to which one pleases others, and doing the technically best thing does not always please others.

That this obvious, inherent conflict of interest is obscured by the popular stereotype and myth is demonstrated in figure 9. Good cartoons and jokes usually work because they surprise us in some way, by emphasizing something incongruous. Evidently, to some people it seems

"Remember, our job is to expand the boundaries of science and make a buck at the same time."

Figure 9. A Frank Cotham cartoon from the *Wall Street Journal,* reproduced by permission of Cartoon Features Syndicate; and with thanks to Dr. Michael Loop, who brought it to the author's attention.

incongruous that scientists do need to earn a living even as they seek
new knowledge.

Conflicts of Interest

That conflict of interest is inherent in being a scientist does not call
for shoulder shrugging; rather, it calls for determination to make the
conflict minimal. In peer review, for example, one ought always to keep
technical criteria preeminent above personal ones to the extent possible;
and the possible extent is greater if one abstains from judging relatives
or friends.

I see great cause for concern in that influential people and organi-
zations nowadays seem oblivious to the need, if science is to be good,
for conflicts of interest to be studiously limited, if not entirely eliminated. My concern is all the greater when simple understanding seems
to be lacking of what conflict of interest is. I have already quoted the
Nobelist who was certain of his integrity even while playing roles *whose
interests are properly different.* Again, in a notorious dispute over flaws in
a published paper coauthored by David Baltimore, the National Institutes of Health set up a panel of three people to investigate: "One had
co-authored a textbook with Baltimore, the other was his co-author on
over a dozen papers. . . . The NIH official responsible for the appointments told the incredulous congressman, 'Sir, I still believe they would
have been objective.' " So deep, apparently, is the faith that scientists
are, or can be, objective to a degree that is quite unnatural for human
beings.

Some of the shoulder shrugging seeks to justify itself under the claim
that not all conflicts of interest actually exist. Thus an official of the
National Association of State Universities and Land-Grant Colleges op-
ined, "Many people have apparent conflicts of interest"; and proposed
guidelines at the Whitehead Institute prescribed that appointments be
made "with consideration of real or apparent conflicts of interest."
Now it is true, of course, that the existence of a conflict of interest
does not inevitably signify that a specific wrong will ensue; but even
where no wrong eventuates, the conflict of interest remains quite real
and not merely "apparent." The idea that conflicts of interest should
be avoided is founded on probabilistic reasoning: given one hundred
decisions to make, a person who may judge well eighty times in absence
of a conflicting interest might judge equally well only seventy times if
there is an appreciably conflicting interest. That one decision, or a year's
record of decisions, or any finite number of decisions reveals no consequence of a conflicting interest does not mean that tomorrow the

conflict of interest may not turn out to be fatally damaging. Experience of human nature teaches that a person who has once resisted temptation may nevertheless succumb on a subsequent occasion.

We know that human beings tend to follow their interests, and when they have conflicting interests, something must go. To avoid conflict of interest is not just to safeguard others who might be affected, it is to shield the individual who is most directly concerned from situations that are fundamentally irresolvable. Every avoidable conflict of interest should be avoided. To speak of "apparent" conflicts of interest is to muddy the waters.

Further, even if no specific wrong eventuates from conflict of interest, a general air of misease comes to prevail. A university's vice-president for research, forced to deal with academics who also run a company that profits from their academic research, found that "the experience . . . taught him an important lesson: the appearance of a conflict of interest can be just as damaging as the real thing."

At Health Stop services, physicians who prescribed clinical tests received a percentage of the profits earned on the tests. An independent study found that physicians prescribed more tests under these circumstances than when they did not share in the profits. The CEO of Health Stop called the study inconclusive. Yet no study, surely, is needed to discern that a conflict of interest exists to which some people may sometimes succumb; and that the conflict is avoidable and therefore should be avoided.

Physicians inevitably suffer conflict of interest between careerism and providing care; an even more severe conflict is inherent in clinical research, between giving care and acquiring knowledge. For valid knowledge to come about, clinical trials must compare those receiving treatment with "controls," similarly ill individuals who are *not* given the experimental treatment. Moreover, the trials must be "double blind," which is to say that even the physicians caring for the patients do not know who is receiving treatment and who the placebo. One who breaches the protocol—say, because the treatment already seems so effective that everyone should benefit from it—vitiates the validity of the experiment: the *seeming* effectiveness may be an illusion, which is why controlled statistical studies are necessary.

An instance where the desire to give treatment outweighed the seeking of knowledge was the providing of the newly proposed drug DDI to any sufferers from AIDS who asked for it; in the absence of controlled trials, we shall never know how many of those who died while taking the drug would have died sooner and how many later had they not taken it. To the other side, human volunteers have been used to acquire knowledge through procedures that offer the volunteers no potential benefit other than to have helped in acquiring the knowledge, as in the labeling of certain cells in terminally ill sufferers from cancer. Most frequently, though, it is not obvious where treatment ends and research begins—say, in the implanting of artificial hearts.

I was personally fortunate to be accepted, in the early 1980s, in the NIH trial of balloon angioplasty for clearing coronary arteries, and I remain grateful for the rejuvenation I experienced. After the (second) angioplasty, I took an experimental drug to lessen the probability that complications would ensue. After several weeks the medication was stopped when certain substances showed up in blood tests. Two years later, after a follow-up examination, one of the researchers asked whether I would be willing to take that medication again: I was one of only a small number of those taking it who had shown the side effect, they wanted to learn more about it, and as part of that they wanted to see what happened when those who had reacted as I did were "re-challenged." I gladly accepted the scrupulously given suggestion that I was perfectly free to refuse—even though, out of gratitude, I wanted to be of help in the research. It seemed to me that the knowledge to be gained was not worth the risk.

Consenting human beings can be experimented with rather readily: as one practitioner said, "When we were in an academic setting, we more or less could make an antibody on the lab bench and inject it into patients next week. . . . an academic situation doesn't have to live up to the guidelines that a big company has." Perhaps all statements of informed consent in clinical research should carry this message: WARN-ING—PARTICIPATION IN THIS TREATMENT COULD BE DANGEROUS TO YOUR HEALTH. YOU WILL BE TREATED BY PEOPLE WHO CARE MORE ABOUT GETTING RELIABLE DATA THAN ABOUT YOUR WELFARE.

The conflict between caring and discovering is difficult enough without introducing directly commercial factors. Some of those studying TPA as treatment against heart-attack damage also held stock in the manufacturer of TPA. An eye medicine was tested by an academic re-searcher who, before publishing the results that the medicine was in-effective, sold his shares in the manufacturer. Little wonder that Congress urged the NIH to draw up regulations governing conflict of interest. Draft guidelines "would have barred N.I.H.-supported scien-tists, research administrators, and their families from owning stock in companies that might be influenced by the scientists' research." A rea-sonable enough proposal, one would think; yet it encountered a storm of protest.

Some clinical researchers argued that "an individual's bias . . . could hardly be a major factor in influencing the outcome of rigorously controlled multi-center trials." But if there are no rules against a conflict of interest and if shoulder shrugging continues, most or all of those doing the "rigorously controlled" trials in all those different centers, not merely a single individual, are likely hoping for a particular result.

Others wondered, "How . . . could you know in advance exactly which companies might express an interest in a basic research project?" Because scientists are smart enough, one might respond; because they do think about possible consequences of their research. Beyond that, if something unforeseen turns up and creates a conflict of interest that was not there before, then one could handle the issue there and then without penalizing anyone for failing to foresee something unforeseeable.

One professor and company owner advocated shoulder shrugging in this way: "If you take away anybody with a conflict of interest, you take away all the experts." One is tempted to respond, "Really? Let's do a survey." The entrepreneur also pointed out that "the alternative is to let the Japanese buy the United States." Again one would like to respond, "How come? Do the Japanese ignore conflicts of interest? Or is it just that Japanese efficiency and quality control is superior?" The entrepreneur's vice-president for research also advocated shoulder shrugging: "Blanket prohibitions don't work," he said. By the same reasoning, we ought to scrap the speed limit, or the blanket prohibition on driving while drunk, or on getting married at age seven.

Another slant on the matter came from the president of the Pharmaceutical Manufacturers Association: "In most instances, private interest and public interest can coincide without interfering with the objective conduct of publicly funded research." Now several decades ago, Secretary of Defense Charles Wilson, former CEO of General Motors, ventured that "what's good for General Motors is good for the country." He was inundated by laughter, sarcasm, and rebuke. Nowadays such statements seem to evoke only shoulder shrugging or head nodding.

"The cruel irony is that these 'Proposed Guidelines' come . . . when Congress, the Administration, and the Public are pushing for a closer cooperation between the NIH, academic research, and companies," pointed out a venture capitalist. But the guidelines say nothing to hinder *proper, ethical* cooperation. It is the case, though, that congressional initiatives intended to stimulate technological developments have blatantly encouraged the countenancing of conflicts of interest: "The U.S.

Technology Transfer Act of 1986 . . . encourages federal grantees and employers to profit from their discoveries."

The point is that these public statements reveal a lack of understanding of conflict of interest. They advocate shoulder shrugging lest worse things befall—in other words, the means are supposed to be excused by the ends.

What needs to be said is that the validity of science is safeguarded by no impersonal, objective method or methodists but by the imperfect knowledge filter. That filter becomes more or less perfect according to the integrity of individual scientists. And it is asking too much of human beings that they always act only in the interest of abstract virtues while they are subject to conflicts of interest.

Support of Science

The filter can work only on what is poured into it. In the early days of science, that was governed very largely by individual human traits of curiosity and by happenstance; only a few students of nature needed help to purchase large telescopes, for example. Nowadays, it is becoming rare for a scientist to be able to do much without specific support. Even at "research" universities, in which doing research is said to be upward of one-third of a professor's responsibilities, professors are expected through entrepreneurship to find the wherewithal to carry out that responsibility. So the progress of science is increasingly influenced by those who control the purse strings.

The degree to which control can or cannot be exercised over science or technology is discussed in chapter 6. Here seems the place, though, to point to conflicts of interest in those who hold the purse strings. For the past several decades, government support for research in the United States was channeled through such agencies as the Atomic Energy Commission, the Department of Defense, the National Science Foundation, and a good many others. The government decided how much to allot and the agencies decided how best to spend it, using technically knowledgeable people to decide where the best value would be received among the available researchers and institutions. In the past few years, however, Congress has increasingly come to include research funds in the pork barrel: bills are passed to establish a supercomputing center *at Cornell University,* say, rather than to give the National Science Foundation funds to establish a supercomputing center wherever the scientific community agrees it would serve the most good.

Pork barreling is an attempt to compromise among conflicting interests: to please one's constituents and to promote scientific progress.

Clearly enough, the likelihood is that such compromises will not benefit equally both of the objectives. Derek de Solla Price has argued with much evidential support that a small percentage of scientists produce almost all the good science; and most scientists would agree that it is largely a waste to support anything but the best-qualified researchers. It is rather dismaying, then, when influential people and groups make their peace with pork barreling instead of pointing to its inconsistency with the ideals and the best practice of science. The Association of American Universities, comprising the few dozen most elite research institutions, refused to criticize those of its members who lobbied for pork, which involves hiring lobbyists at six-figure fees and more: "the informal but official position of the academic community's elite is that pork barreling, although naughty, is no longer so damnable as it once was. . . . The reason is that the facilities situation is felt to be critical." The ends, in other words, are taken to excuse the means.

Scientific Communities

Criticism from colleagues or peers compensates for the human inability to be skeptical about one's own beliefs and to be aware of one's own biases. But if one belongs to a rather homogeneous community, particularly if it is a small one, then the potential critics may share one's own beliefs and biases. Whole groups of people may then fall into error for longer or shorter periods of time.

Homogeneity and isolation may exist for reasons of scientific history or for reasons outside science. It is well known that scientific specialization involves a degree of isolation: separate societies and journals are founded to cater to increasingly specialized groups. But isolation can stem from other circumstances too: science in many countries has not shared effectively in the state-of-the-art consensus, through lack of opportunities for advanced training or for access to modern equipment, or as a result of language barriers and lack of up-to-date literature, or through politically enforced isolation. No matter what the origin of isolation or homogeneity, however, hindsight makes plain that error flourishes thereby and that stagnation is more likely than advancement.

In new subspecialties, or in old but small ones, a consensus may well be less than judicious, a fad or even a communal version of *folie à deux,* not much informed by reality therapy. So long as they do not need to explain themselves to outsiders, people can maintain views of the most fragile validity, about science as much as about politics or religion. But organized science nowadays comprises largely *overlapping* communities, and eventually consensus over the most salient things must be shared by all—they are all working the same jigsaw puzzle, albeit different parts of it.

One sort of lapse can occur in geographically isolated communities. A standard example is the purported discovery of N-rays by René Blondlot and his colleagues in Nancy, France. The measurements depended too much on individual judgment; but so far had the studies been taken in isolation that, by the time outsiders were able to comment, Blondlot had become too much set in his beliefs to be able to benefit from adverse criticism.

Defects can also permeate a community that is parochial technically rather than geographically, so that useful knowledge existing in other fields is not drawn upon. A possible example here is in electroanalytical chemistry. In the 1960s, someone in that working community learned about operational amplifiers, which lend themselves as building blocks for more complicated electronic circuitry; and for a decade or more, circuits were designed and published and used that were inferior to what electronic engineers could produce.

Groups may be isolated by fiat rather than happenstance. Industries train their scientists to listen rather than to talk. At his invitation, I once visited an electrochemist at his industrial lab and was shown the facilities. When I asked, "And what are you working on?" I was told, "We're doing some reductions; and also some oxidations." I waited in vain to hear anything more, anything that we could usefully have talked about. I do not question the need for patents and profits, but the dangers of working in effective isolation can be considerable. For example, the desire to withhold detailed crystallographic data recently led at least one industrial group to work in a fallacious direction; had they published their data, others would have found the error for them quite quickly.

Secrecy imposed for supposed reasons of national security may well damage rather than enhance that security. Scientists like to remember Robert Oppenheimer's achievement in managing the development of atomic bombs in part because he insisted on a high level of open discussion among the various technical groups at Los Alamos. Oppenheimer understood, as his military counterpart did not, that free-flowing criticism from people with different viewpoints could efficiently expose error and also generate unforeseeably useful ideas. It would be interesting to see how proposed new systems of weapons would fare nowadays if they were subject to review by the whole technical community. I would expect reviewers to recommend against the Strategic Defense Initiative, pending demonstrations that sufficiently powerful lasers are feasible, computing needs manageable, and the whole system robust enough to withstand attack; against the Stealth bomber, pending demonstrations that flying wings are not inherently unstable; and so on.

The dangers of isolation or secrecy are clear. As Peter Medawar has remarked, those who shut their doors *keep* out more than they *let* out.

Historians and sociologists of science have noted many instances of discoveries made by mavericks or by people who moved from one specialty into another—people who were able to see things hidden from those steeped in the old consensus. When isolation is broken, fresh viewpoints can be remarkably illuminating. So too do experienced teachers find that students who ask naively uninformed questions can sometimes expose areas of ignorance studiously avoided by the practicing experts; for example, as I was trying to learn electrochemistry I discovered that electrochemists for several decades had used a formula based on no agreed theoretical justification and little empirical warrant.

Bias and Progress in Science

Science can quickly be brought to a standstill when it is forced into the mold of an ideology. Physics could not flourish in Nazi Germany when it was commanded to do without the "non-Aryan" theory of relativity. Soviet physics suffered similarly, as did Soviet chemistry when commanded to do without wave mechanics. The best-known example is that of Soviet genetics, which could not be carried on scientifically for decades while the incompetent Trofim Lysenko was politically empowered. Those episodes exemplify evasion of reality therapy, not the practice of science.

It is not only under repressive political regimes that science may be hindered by ideology. Presently in the United States the media and funding agencies discourage those who want to study the heritability of intellectual qualities; and no government funding is available for research on several aspects of human reproduction. Research that appears consonant with prevailing attitudes is well publicized; not so when it gores established cows—for example, the finding that fluoride may be associated with certain cancers.

Science has made undeniable, often rapid progress without noticeable interruption since the seventeenth century. But no law requires that progress will always continue: that is up to the puzzlers and the social environment in which they play. There must be players who want strongly enough to make the puzzle grow—strongly enough that they submit to reality therapy; and their society must permit them to play openly. Otherwise, parts of science can come to a stop and decay: as

Greek science ceased many centuries ago, and Islamic science later, and Nazi and Soviet physics for a time, as already noted.

To assure that scientific knowledge is reliable, and that progress is rapid, requires that interactions among scientists be unconstrained and that scientists be as varied as possible in their biases. Science progresses through continual winnowing under consensually governed institutions. Objectivity comes into science because ideas and results are exposed to the criticism of people with disparate and conflicting and competing intellectual approaches and beliefs, personal biases, social goals, hidden agendas, and the rest, so that—by and large—consensus among all of them can only be achieved when they are left no other option than to agree with each other, when the puzzle itself demands and allows it, when the players submit jointly to reality therapy. Scientific activity therefore becomes more efficient and more reliable the more it includes the whole range of human types—geographic, sexual, intellectual, emotional—just so long as they want to learn about nature and are willing to endure the stress of reality therapy.

Here is the soundest argument for affirmative action: it is good for science itself. Most of the contemporary calls for increased participation in science by women and other traditionally uninvolved groups neglect to make this point as they rely on social ideology, on notions of justice for those groups, or on vague, unsupported references to "alternative culture" or "alternative science." An understanding of how science actually works demonstrates that it can only be to the benefit of science itself if scientists are drawn from the greatest possible variety of human types.

That science is better the less it is constrained stands in direct opposition to those who want it to be conditioned or led or commanded by Marxism or feminism or New Age notions or any other set of prior beliefs. Calls for ideological control are based on the misconception that the world can be tailored to our liking, that knowledge about the world can always be made, if not desirable, then at least palatable.

6

Consequences of Misconception

When the nature of science is misconceived, inevitably the influence of science on practical affairs is also misconceived.

Frontier Science and Textbook Science

Textbook science is uncontroversial. It seems to tell of a clear and steady progression through the centuries, from the times when little was known to the marvelous cumulation that we now control. We all learn science from textbooks, and we can hardly fail to be impressed by the range and reliability of the knowledge that has by now been amassed and by the power of the theories that orchestrate that knowledge.

Though we are impressed, we rarely remain excited by this, at least not for long. What we do get excited about is the very latest stuff: the wobbling of a star that might be the first demonstration of a planet around some other sun than our own; or the synthesis of a human gene in a test tube; or nuclear fusion at room temperature in an electrochemical cell.

What we get excited about is also what gets into the newspapers and magazines and onto television. But what we mostly fail to realize is that what we get excited about—namely, the frontier stuff—is quite a different sort of science from the textbook variety about which we learned in school and college. Frontier science is very *un*reliable. The wobbling star soon turns out not to be wobbling at all; or it turns out to have a dark companion, so that it is just another example of a double star and not the indubitable discovery of the first planet outside our solar system. The "fastest-spinning object in the universe" turns out instead to be an ordinary TV set interfering with the radio signals studied by the astronomers. The synthesized gene turns out not to work when it is put into a living system. And so on. Most of the miraculous science that gets headlines today fades away relatively unnoticed by next week or next month or next year, as it has to run the gauntlet of the scientific community, as others try to use the result or to repeat it.

The trouble is that we use the word "science" to describe both the reliable textbook stuff and the exciting frontier stuff. So we fall into the very bad habit of taking the frontier stuff to be as reliable as the textbook stuff, and that has some quite debilitating consequences.

Fraud

There has been much fuss in recent years about fraud in science. Newspapers and magazines and congressional committees have all declaimed a positive epidemic of fraud and that something must be done about it—some way must be found to avoid fraud occurring in science. The intensity of concern shows how science has become for our society the supreme arbiter of what is truthful and what is not truthful. Fraud in science thereby becomes treason to the whole society, even to all human societies, with incalculably serious effects.

That view, however, is based on misconception, most directly on the failure to distinguish frontier science from textbook science. Almost all the allegations of fraud have involved rather recently published—or even not-yet-published—science—that is to say, frontier science. Now no one ought to place much reliance on frontier science under any circumstances. It is inherently tentative, controversial, biased, unreliable . . . ; indeed, it lacks all the virtues for which we prize and admire textbook science. Just as innocent error is filtered out as frontier science matures in time and through the workings of scientific institutions, so deliberate error too is weeded out, by the same processes and for the same reason: false science cannot persist just as unrepresentative maps do not continue to be used—the discrepancies with reality get noticed.

One might truly begin to worry if one were to find such a thing as the Ideal Gas Law or the Periodic Law of the Elements to be both wrong and fraudulently developed. But nothing like that has happened; nor will it happen, because any such *important* piece of science has been used, and thereby continually tested, innumerable times by different people under all sorts of circumstances *before it is acknowledged to be important and becomes relied upon.* Fraud can long remain undetected only where it concerns something recondite, of little general interest. But then its consequences are not great, for it will not have found a place in the textbooks and will have remained simply a report of research, which no sensible scientist places too much reliance on in any case.

Some iconoclasts have leveled accusations of fraud even at some of the greatest scientists of the past, at Ptolemy and Newton and Mendel. But the allegations are that these people did not scrupulously follow the scientific method and its ethical corollaries, not that Newton's laws of mechanics or Mendel's laws of heredity are wrong. And here is another benefit of exposing the scientific method as myth: so long as that myth is believed, one can be led to describe as misconducted some of the greatest advances in knowledge ever achieved by human beings! But if it is recognized that the scientific method works communally, not for each individual scientist, then one remains free to admire Ptolemy, Newton, Mendel, and the other great people who undeniably made undeniably great contributions—albeit they were human beings, not gods or disembodied intelligences. Surely it is somehow inappropriate, if not bizarre or perverse, for pundits who contribute nothing themselves to accuse accomplished people of misconduct by measuring their behavior against a purely mythical standard.

The essence of the charges is that numbers calculated from theory were passed off by Ptolemy as actual measurements, and that Newton and Mendel "trimmed" the measured data to make them more convincingly consonant with their theories. Certainly that is wrong; any scientist who does that deserves to be drummed out of the profession. But there are some caveats especially where the past is concerned. First, as already discussed, scientists must judge what to publish, and one cannot at once conclude that a person was cheating—rather than exercising poor judgment—in selecting what to make public. As also already noted, refusal to make the jump from experimental data to its meaning in terms of rounded-off numbers can vitiate or delay understanding. Second, one cannot assume that the relation between data and theory was always viewed just as we view it now, namely, that ultimately observation should have the last word. Certainly there was no such widespread belief in Ptolemy's time. As to Newton, we do not find it easy to understand one who is acknowledged a great scientist and yet who also spent a great deal of time on alchemy and biblical exegesis; and Mendel's mind, too, is not quite an open book for us. Perhaps he viewed the numbers of pea plants he could count as in the nature of a parable, not important in itself, not in a strictly literal way. It is difficult enough to prove a living person guilty of *deliberate* deceit; even greater caution is appropriate in finding guilty those who came before.

But what about more recent instances, one might ask, like the Piltdown fraud, which hampered paleoanthropology for decades? What about the psychologist whose fraud resulted in inappropriate medication of countless retarded or hyperactive children? Surely tangible harm was done in these and in similar cases; surely that ought to be taken seriously; surely something ought to be done to prevent that sort of thing from happening again?

First we need to be clear about what the problem is; only then can we fruitfully consider what can be done about it. In attempting to comprehend the problem, it needs to be understood what science is and what it is not. Among other things, it is vital to distinguish among the various sorts of science, for what is true for one is not necessarily true for the other.

Piltdown Man, though not exposed as fraudulent for about four decades, nevertheless had hardly become a piece of textbook science in that time: it had remained in a sort of limbo precisely because it did not mesh with other relevant knowledge. Indeed, paleoanthropology as a whole remains largely a frontier science, and it is also a data-poor science. It advances at a snail's pace compared to chemistry, say, and a thing can remain unreliable frontier stuff for decades. The solution, then, is not to attempt the impossible—namely, to prevent any prospective hoaxers from leaving pseudofossils for others to find—but to be clear about what makes any given piece of science reliable or unreliable, and to what manner of degree and under what circumstances.

So too with newly proposed medications. Suggestions are always frontier, not textbook, science; and they may be seriously wrong, equally for innocent reasons as by reason of deceit. There is simply no way to ensure that proposed treatments—hunches, inspired guesses, good ideas—will always be effective; there is no way to ensure that they will never be harmful as well as ineffective. Unless we are willing to settle for what we have now, we must try new things, and anything new carries a risk. The only possible solution is continuing caution and proper clinical trials of progressively larger scope and greater duration, so long as no ill effects are observed. Who is really guilty if caution and common sense and understanding of scientific activity do not prevail and some newly touted remedy is embraced at face value? Wishful thinking is the greatest part of it: we so much want to be cured—or to be able to cure others—that we forget to look carefully at whether the cure has been proven. Hence Christian Science, homeopathy, and many other alternative systems, as well as innumerable individual panaceas such as monkey-gland therapy or laetrile. If we believe what others say without proof, the consequences are likely to be unfortunate, be it on a matter of medicine or of science or of politics or of economics.

That we should like to do a thing does not make that thing possible. That we want science to be true cannot make frontier science reliable. But the myth of the scientific method does not call for us to distinguish between reliable textbook science and unreliable frontier science, and so we do not even recognize that it is our own wishful thinking that leaves us open to quackery and fraud. And because we hold dear the myth of the method and its attendant misconceptions, we take inappropriate measures against fraud in science. Thus there is established, in the federal bureaucracy, an Office of Scientific Integrity—surely by people who never read George Orwell's *1984*. No matter what explicit and reasonable-sounding justification may now be given for it, the arguments that led to this step show plainly enough what the ambition was: that no one be misled by something fraudulent in science, that cheating be exposed *just as soon as it occurs*. Which, on reflection, ought to be seen as plainly impossible. Quite in general, attempts to do the impossible tend to do harm. Here, one danger is that in order to detect fraud quickly, suspicion of one's colleagues, students, and supervisors must be ever present, an atmosphere hardly conducive to enthusiastic teamwork. Traditionally, one assumed that all was well until *forced* to admit the possibility of deceit—but that takes time. (Not necessarily so very long, though. Most of the misconduct about which the fuss is being made was noticed and decried within months, often before anything had even been published.)

Again, note that fraud in science seems so immediate and horrendous a problem only because we misunderstand science and misunderstand the ramifications that fraud can have. If we discard the myth of the scientific method and recognize science as the complex human activity that it is, then we can deal with fraud in science just as we deal with fraud in business, government, or anywhere else: unhysterically, under the rules we have evolved that safeguard the accused until such time as guilt may have been proven beyond reasonable doubt. But, again, we don't wish to discard the myth. We *want* to believe that science, be it frontier or whatever, can give us what we wish. We *want* to believe that the cure for cancer is just around the corner and that we will get to it faster the more money we spend. We *want* to believe that we can buy whatever science we have a mind to. And so we continue to fool ourselves about science. As the victims of confidence tricksters actually are not fooled by others but are simply given the opportunity to fool themselves through greed and wishful thinking, so our society, greedy for what it thinks it can get from science, fools itself—and then waxes hysterical when some part of the fable is exposed, that one of our scientists happens only to be wearing emperor's clothes, for example.

Society has no need to be as hysterically fearful about fraud in science as it is now fashionable to be. It is extremely unlikely that a fraudulent claim about an important scientific issue will cause much harm, because as others strive to capitalize on that important claim, they find themselves unable to do so, and the fraud is exposed or the claim consigned to limbo.

But to advocate reason as against hysteria is not to say that attempted frauds in science have no consequences at all. People's time is wasted following a false trail; more seriously, the necessary atmosphere of trust among scientists is eroded; and the danger to science as a profession is considerable. The recently fragile credibility of the scientific community gets more fragile, yet many scientists fail to acknowledge that fact because they, too, believe the myth of the method. Under that myth, science is *automatically* self-correcting, individuals who cheat are individual aberrants, *and therefore the scientific community need do nothing;* there is no fault in the system *because there cannot be—it is an impersonal system, not at the mercy of human vagary.*

In reality, of course, as the filter and puzzle model makes plain, the efficiency with which science works depends on the competence and integrity of the scientific community in evaluating its members, in judging grant proposals and manuscripts and job seekers and nominees for prizes. The system works better the more each member of it strives to behave according to the ideals expressed in the myth of the scientific method and the Mertonian norms. That textbook science now is relatively independent of human vagary gives no assurance that this or later generations will be able to add to that knowledge. The history of science tells of self-deception by individual scientists and by whole groups of scientists; it tells of the routine rejection by the scientific community of brilliant ideas that sometimes turned out to be worthwhile; it tells of instances when the state commanded science to go astray and some scientists became willing Quislings while others shrugged their shoulders and went about their business as defined for them by the state.

The history of science gives no grounds for the scientific community to believe that science will always and automatically be allowed to go its own self-governing, reality-seeking way. Rather, the lesson of history is the same as for all human beings, and in particular for all professions: the price of liberty is eternal vigilance. One can go irretrievably wrong through a long enough series of harmless-appearing little steps. Professions that do not keep themselves ethical and credible lose their autonomy.

Medicine

The distinction between frontier science and textbook science is crucial in medicine at all times, not just under circumstances of fraud or alleged misconduct. The search for new medicines offers many contemporary instances in which wishful thinking ignores that frontier science is not reliable knowledge. Over many years, a conservative protocol has evolved to screen substances before they are permitted into general use by human beings. Drug companies and others periodically have complained about it, pointing out that new drugs could be obtained in other countries years before they became available in the United States. But such complaints fail to take account of the inherent fallibility of the relevant knowledge. Understanding the progression from frontier science to textbook science enables one to understand that newly introduced drugs are inevitably risky; that tests and trials make them progressively less risky—for those substances that pass them, that is to say; but that the risk never reaches zero. Urging that new drugs be approved more quickly is the same as urging that higher levels of risk be accepted; but that is not what is said (or even believed).

In the hysteria over the AIDS epidemic, various groups even claimed conspiracy to withhold treatment because the United States did not permit the sale of putatively useful drugs before their safety and efficacy had been established to the traditional level. Under pressure, the regulations have been loosened. Whether that brings beneficial treatment nearer is a moot point, however, and in all likelihood will remain moot: unless clinical trials are conducted, no accurate knowledge of efficacy or harm can be obtained, and if everyone is free to try everything, clinical trials are not possible. What is not moot is that loosened regulations make it easier to use human beings as experimental creatures, as the proverbial guinea pigs. In a number of ways, in fact, it is now easier to do experiments on human beings than on other animals: the regulations governing the use of animals is very strict, but clinicians have much leeway with human volunteers, especially with those who seem in any case to be on the verge of dying. (One rather trivial and therefore amusing example: a number of years ago, at the University of Kentucky but at the behest of federal regulations governing the welfare of animals, the pigeons and rats housed in the old, not air-conditioned psychology department had to be moved into the air-conditioned chemistry department; but the human members of the psychology department were not eligible for such relief.)

As earlier discussed, clinical research carries a built-in conflict of interest between caring for individuals and acquiring valid new knowledge. Similarly, in the practice of medicine hoped-for benefits have to be gauged according to where the available treatments lie on the range between frontier science and textbook science. Physicians ought to be clear in their own minds about that. What those who complain about conspiracy and long delays in approving drugs do not usually mention is the classic case of thalidomide, the tranquilizer used widely in other parts of the world before it had been approved in the United States and which turned out to cause serious birth defects when taken by pregnant women. Only the phlegmatic pace of the American system of approval saved tens of thousands of American babies from being born without usable limbs.

Those who complain that treatments are available in other countries but not here should also know that much quackery is also available in other countries. In the Philippines and in Brazil, for example, spirit healers and psychic surgeons openly ply their trade. In Britain, alternative medicine is becoming more and more visible, and one can have one's illness diagnosed by iridology (looking at the iris of the eye), or by reflexology (examining the sole of the foot), or by other occult means, and one may then be treated by the administration of remedies that have been diluted so far as to be composed of the purest of pure water (homeopathy).

We simply do not now know how to cure many of the ills that humankind is prone to. There was wisdom in Hippocrates when he made the first rule of medical treatment, "Do no harm"—not even, or perhaps especially not, inadvertent harm through wishful thinking. For as far as we can look into the future, much of the practice of medicine—as also public policy-making—will involve the making of decisions without having all the knowledge that one would like to have; and the less reliable the knowledge, the greater the risk.

Scholarship about Science

Some long-running disputes among historians, philosophers, and sociologists will be resolved only by recognizing the distinction between frontier and textbook sorts of science. Traditionally, there have been two schools of thought about the historical development of science: the internalist school and the externalist. According to internalists, since correct science reflects nature, it is unaffected by the human traits of ambition, ideology, prejudice, or dishonesty that individual scientists might have displayed or the particularities of the human societies in which they lived. Thus internalist history is purely intellectual history, tracing the development of scientific theories toward those now held to be true. According to externalists, by contrast, the science that people produce, just as everything else they do, reflects their biases and wishes and social environment; and so externalists are as much interested in dead ends and failures in science as in successes.

Historians and philosophers of science were predominantly internalists until a few decades ago. Their focus was on the successes and growth of science—in other words, on the logic and power of *textbook* science. As earlier described, that focus led naturally to explanation in terms of the scientific method; and anything that did not fit with the notion of a logically impartial method was overlooked or brushed aside. Historians could hardly avoid noticing that some brilliant discoveries reflected passionately creative inspiration rather than careful observation and dispassionately judicious choice of hypothesis; but the import of this was discounted as philosophers of science argued for a distinction between "the context of discovery" and "the context of justification." The context of discovery included the human and social characteristics that were regarded as not amenable to logical analysis: genius, serendipity, and the like. Since those belonged, it was said, to the nonrational part of human experience, they were by definition irrelevant to science, which was the preeminently rational enterprise. Only the context of justification was supposed to bear on the nature and progress of science, in which the ideas that came mysteriously from somewhere or other were selected out logically and impartially by the scientific method. This argument is circular, of course, but that went unnoticed so long as the conclusion (that is, the assumption) was widely enough shared.

This view soon met explicit challenges, however, most fatally in Thomas Kuhn's demonstration that the actual practice of science does not illustrate application of the scientific method. Seizing upon that, some jumped to the extreme opposite conclusion of the relativist view. Since history shows all scientists to have been quite fallibly human, went their argument, therefore science—the things done by scientists—cannot validly claim to be any more objective than literary criticism, art, the law, or any other human enterprise. Science's claim to true knowledge about the world is therefore spurious, continued the critique; it serves the interest of the scientific community, which is allied with the political and social establishment, to maintain the illusion of objectivity in order to maintain intellectual authority. Just as all people acquire the beliefs that serve their interests, runs the externalist, relativist analysis, so science inevitably develops the knowledge that serves the interests of the society in which it is nurtured. So, for instance, Darwinist theory of the survival of the fittest suited the purposes of laissez-faire capitalism and emerged for that reason in that time and place, not because the idea has intrinsic validity.

Curious as it may seem in view of the radical implausibility of both extreme positions, argument still continues between the two camps; not always explicitly, to be sure, but nevertheless clearly enough since some scholars remain congenitally internalist and others congenitally externalist. (Sociopolitical arguments are analogous, in that everyone agrees *in principle* that a combination of heredity and environment makes a person; but in practice some people focus incessantly on hereditary factors whereas others study and talk of nothing but environmental influences.) Thus philosophers find it difficult to incorporate human waywardness into their professional discourse and sociologists resist the idea that the actual behavior of some human beings may be significantly influenced by logic and selflessness. Marxists (among others) continue to press an extreme externalist view, never mind that they ought to have learned a lesson from the Lysenko affair (among much else); extreme feminists continue to insist that scientific knowledge is somehow masculine and power hungry; third-world ideologues characterize science as imperialist and Western.

In reality and quite obviously, internalist and externalist positions are both right and both wrong, because some bits or aspects of science lend themselves to internalist explanation whereas others yield naturally to externalist exegesis. Textbook science is best understood through an internalist approach that illuminates the intellectual history of the growth of ever more reliable knowledge. Externalist criticism is pointless here unless it could be shown that, say, a Marxist, feminist, or third-world science could incorporate a different Periodic Table of the Elements (a nonperiodic one, that is to say), or some other equation than $E = mc^2$, or some nonentropic system of thermodynamics, or something equally beyond the bounds even of science fiction. Frontier science, on the other hand, is aberrant, paradoxical, and discomfiting to internalists because logic and coherence and reliability are so little in evidence; the frontier becomes fruitful under externalist examination, which is interested, for instance, in error, in why some errors were perpetrated rather than others, in why some errors persisted and others did not, and so forth, and which does not focus exclusively—as internalists are wont to do—on what ultimately stood the test of long time.

The internalist-externalist distinction became an issue first in history, because historians are trained primarily to trace the details of a story. Given the intricacy of human affairs, no one story can ever do justice to all detail, and individual historians had long been allowed much freedom in choosing to emphasize some features over others. So some historians focused on intellectual issues and others on social institutions, for example; and eventually the community of historians could not help but notice an increasing conflict of explanation between different schools, and to look for resolutions, and thus to make explicit the internalist-externalist dichotomy.

Philosophy, as indicated earlier, is by nature internalist, whereas sociology is externalist. Scientists themselves are steeped in the myth of the scientific method and, having learned initially from the textbooks, they become instinctively internalist. Many scientists would agree with Fred Hoyle that "science is not made by what communities think, but by what the Universe is"; and when one looks only to the long run, to textbook science and not to the frontier, it is easy to slip into that opinion.

Those who aim to measure scientific literacy would do well to recognize the frontier/textbook distinction. Do scientific literates need to be familiar only with such well-established concepts as the chemists' notions of atoms, or must they be cognizant as well of the latest notions about big bangs and black holes? If the latter, then such contradictions will emerge as earlier discussed, about the possible reality of extraterrestrial intelligence and of an extraterrestrial origin of some UFOs.

Misconceptions and disciplinary discordances make evident the need for interdisciplinary scholarship if the nature and cultural import of science are to be satisfactorily understood; that is why science and technology studies has come into being. The degree to which disciplinary bias can hamper understanding of what science is may be illustrated by the response of the philosophy of science, inherently internalist, to the proposition that science is a human activity. That is the crux of what Kuhn was saying, that the actual actions of scientists need to be understood and that those actions are not explicable as purely intellectual initiatives; yet philosophers of science reacted quite immoderately and chose to attack Kuhn's weaknesses rather than to acknowledge the strength and importance of his chief contribution—for they were only equipped to handle abstract ideas. Recently, as the philosopher David Hull worked through consequences of the fact that ideas can only be carried and transmitted by human beings, philosophers (and others) were again or still unwilling to concede that science has significant truck with emotion and illogic, that human competitiveness might even be one of the motive forces driving scientific activity.

At any rate, without going further into illustrative detail here, it is plain that scholarly attempts to understand science and its relation to culture continue to be hindered by disciplinary narrowness and by the associated tendency for people to be dominantly *either* externalist or internalist toward science as a whole, instead of recognizing the necessary complementarity of those approaches in dealing with science over the whole range from frontier to textbook. Science begins by chance and caprice, at the frontier, with hardly a shadow of the scientific method in evidence; and then it proceeds to be sieved, tested, and modified until it appears in the textbooks, whose tried and refined content can be explained so simply (albeit fallaciously) as resulting from application of the method.

Science in the News

Magazines, newspapers, and television have little interest in textbook science; only the newest and latest is their grist. What they cover is frontier science, capriciously unreliable and fraught with often unsavory "personal-interest" attributes; but they cover it under the label "science," which to them, as to the rest of us, connotes objectivity, reliability, and the scientific method. Thereby confusion is worse confounded, especially since the media's striving to cover *all* the "news" is not matched by any concern to follow up stories covered in the past, even the very recent past. Thus the public is left with stronger impressions of breakthrough, crisis, or drama than of judicious resolution; with stronger impressions of the dangers of pesticides than of the strict regulation that controls their application; with the notion that room-temperature superconductors are in the works, not with the realization that the surge of progress to higher temperatures came to a halt more than 150° short of room temperature. Just as over other matters, so too in science: the media focus on aberration, not on context, perspective, or what is normal.

Now on many other subjects the audience understands what is going on and can make conscious or unconscious allowances out of common sense and human experience: because one president lies, not all presidents are thought to; because someone's neighbor turned out to be a mass murderer, not every neighbor is suspected of being one. But on matters of science, most of the audience lacks the personal experience and technical background to make allowances for the media's bias toward instancy. Since in any case the distinction between frontier and textbook science is so rarely grasped, the media's blurring of that distinction must add tellingly to the general confusion.

Scientists are anything but clear that the media are interested only in the frontier. Being themselves steeped in textbook science, they can hardly imagine what it must be like to lack that background and yet attempt—as journalists must—to comprehend what is happening at the frontier. So those scientists who are reported by the media are often shocked, to some degree or other, by silly errors and lack of context in the ensuing stories; and also by the lack of subsequent continuity of coverage. Those who achieved superconductivity at 100° K and thereby attained instant fame, for example, have also seen how instant is the loss of fame as soon as one stops creating *new* news: emphatic were their protests when a journalist asked, "Superconductivity: Is the Party

Over?" With science now of so much public concern, scientists alike with actors, politicians, and rock singers can appreciate Andy Warhol's insight, that in our fast-track mass society everyone has a turn at being famous, albeit for fifteen minutes only.

Much public controversy about technical issues could be more soundly productive were it generally realized that what is most new is also the least likely to be true.

Public Policy

The myth of the scientific method and the failure to distinguish frontier science from textbook science have their most consequential practical effect in governmental actions. I have already pointed out that the hysterical reaction to a few fraudulent acts by a few scientists stems at least partly from confusing the frontier with the textbook, with Science. Another illustration already given is the drive to have new drugs approved almost as soon as they have been conceived: anything scientific—in other words, anything that comes from a scientist or from an institution that has to do with science—is at once granted the trust (at least initially) that should be reserved for textbook science.

Under the myth of the scientific method, a scientist's advice comes to be seen as objective and to be relied upon. Thus Edward Teller's personal beliefs—because he is eminent, of course, as well as a scientist—come to justify the expenditure of billions of dollars for a Star Wars defense system that most of the scientific community believes to be unworkable. The needed distinction in this case, of course, is not only that between science and frontier science but also that between science and technology, for Star Wars needs not only devices that go beyond current scientific knowledge but also a technology whose complexity, speed, and infallibility exceed anything that has ever been made to work. So too with the Superconducting Super-Collider, the planned atom smasher estimated to cost at first about $5 billion and soon afterward between $14 billion and $16 billion. Because these incredibly expensive schemes come from scientists, normally hardheaded business types and cost-conscious politicians seemingly lose all sense of proportion and the ability to consider potential benefits in terms of potential cost.

Again, within weeks of the announced claim that nuclear fusion had been harnessed "cold" in an electrochemical tube, the governor of Utah had pledged $5 million for a Cold Fusion Center, and a congressional committee had held hearings in which it listened to requests for as much as $125 million. How much public money would that governor have been willing to wager on a horse race? backing even a 50:50 favorite, let alone a 100:1 outsider? Yet at the time the governor made his pledge, that was about the range of odds being offered around the scientific community. Surely any congressional committee would have thrown out on their proverbial ears any group that asked for tens of millions of dollars as stakes in a gamble; yet they heard such requests politely when they came masked under the label "science"—they did not realize that label to be seriously misleading if not downright fraudulent for what is *frontier* science. Only because public and media and Congress and pundits of science do not understand that frontier science is not Science could such a bandwagon boom so loudly and so quickly.

Many misconceptions about science have deleterious consequences in government action. The most damaging are probably the failure to distinguish frontier from textbook science and the confusion between science and technology.

Science and Technology

Confusion is common between frontier science and textbook science because the very basis for the distinction is unfamiliar. By contrast, a distinction between pure science and applied science is widely recognized; not commonly understood here is what precisely the differences are and what follows from them. The most portentous consequence of this lack of understanding is the ill-advised allotting of funds for research: the use of novelty-stifling mechanisms when what one purports to do is stimulate novelty, for example; or such costly and ineffective ventures as the "war on cancer."

Basic Research

The great revolutions of the past frame our picture of science. We honor the great revolutionaries, and to emulate them is our highest aim. So every scientist dreams of making revolutionary discoveries, and the National Science Foundation seeks to stimulate and support such creative, innovative pursuit.

But paradox and self-contradiction abound here. The fundamental law of the unknown unknown is that it is unsuspected, and therefore it cannot be directly explored: scientific revolutions are serendipitous; they come by luck more than by foresight. Successful scientists are typically those who work in the most interesting but still tractable parts of the known unknown; and because they do, it is predictable that they will get results generally agreed to be useful—though one cannot predict how useful or in precisely what way, and one certainly must not expect them to be revolutionary.

Direct, deliberate attempts to uncover the unknown unknown, where the greatest potential novelty lies, are not common in science. How could they be? That would involve following hunches that might lead nowhere at all. Since most scientists (as many other people) are judged, paid, promoted (or dismissed) according largely to what they accomplish, it makes sense for them to choose research that, it can reasonably be predicted, will produce some sort of reasonably useful results. In any case, only a few scientists—chiefly those in universities, but by no means all of them—are free to choose what they will study. Most scientists work in industry or government, under research programs with definite goals within the known unknown. It is a common misconception that science is the search for brand-new knowledge; it is rather the attempt to extend the reach of the known by building on and from it, along lines that the conventional wisdom recognizes as potentially fruitful.

So scientists, individually or in organized groups, do not try to jump directly toward the unknown unknown. They eschew such very-long-shot gambles as looking into UFOs, parapsychology, or cryptozoology (the search for Bigfoot, Loch Ness monsters, and the like). When scientists overtly claim to be seeking novelty, and even as they believe what they say, they define novelty in a limited way, namely, within the conventional paradigm. Novelty is pursued *within the known unknown;* scientists seek what is new, but not what is so new that it could overturn their beliefs.

The National Science Foundation, too, overtly claims to be seeking novelty. But to receive its support, one must make a proposal that specifies in considerable detail what one intends to do, and by what means, and what one expects to find, and what the significance of that will be: one must stick, in other words, to the known unknown. Moreover, one's proposal is submitted to peer review: others in the same field are asked to judge whether the proposal makes sense, whether its aims are feasible ones. Such peer review cannot but be a wet blanket of conventional wisdom, and the private folklore of science understands that the most original proposals are also the ones that find it hardest to get support.

One rarely noted aspect of peer review is that, by and large and on the average but especially with the most brilliant ideas, the reviewers are less qualified than the authors of the research proposals. For one thing, each proposal is reviewed by as many as half a dozen peers, and their average competence is, *solely for that reason,* likely to be lower than that of the author of the proposal: there are fewer brilliant people than there are competent ones, and there are fewer very competent ones than there are moderately competent ones, in any human activity. Further, the best scientists are also those whose time is most in demand and who will not be able to respond to all the requests made to them to review ideas, proposals, papers, books, and so on; and so the burden of doing the peer reviewing trickles down toward those who have more time but less talent. In addition, of course, whoever has evolved a proposal is likely—precisely for that reason—to know more about the specific details of that particular problem than anybody else in the world.

So even moderately successful scientists learn to adjust to the predictability and mediocrity of peer review by camouflaging their best ideas: they seek support for "normal" research into the known unknown but then use some of the granted funds to follow their pet hunches. Those who decided to look at the possibility of cold fusion, though they were accomplished people of high reputation, knew better than to ask for support for their long shot.

Though it is fairly generally understood within the scientific community, it is not usually admitted in public that the grant-allocating mechanism supposed to serve creativity cannot and does not do so. The silence stems partly from the fact that the essential role of serendipity is not fully acknowledged; but partly too it stems from a reluctance, on the part of those who hope for funds, to criticize the people from whom they hope to receive support. So surveys made by the National Science Foundation of those with whom it deals report predominant satisfaction with the Foundation's activities, whereas private gossip among scientists features horror stories of the stifling of originality and creativity. In the mid-1970s a survey of NSF reviewers and grant applicants showed that "they think the peer review system is an appropriate mechanism for making funding decisions, that it works quite well and needs few, if any, changes"; the "one change . . . recommended by almost all applicants involves . . . procedures for appealing adverse . . . decisions." But when the questions were directly about stimulating novel research, "about two-thirds . . . agreed . . . that NSF is unlikely to fund high-risk, innovative research projects because its review process is too conservative. These results . . . came as something of a shock, amounting to 'a serious accusation.' "

That the NSF and its director should be surprised by this is merely another illustration of how widespread are misconceptions about how science works, even among those who manage it and fund it and make public policy about it: that scientific institutions and the scientific striving for consensus tend to stifle creativity hardly ought to be surprising. It is not as though this particular point is an arcane one, after all; as regards unsettling novelty, science behaves just like human societies in general (see chap. 4). Prevailing beliefs, conventional wisdom, paradigms are always inherently conservative; they can never be otherwise, because they can only be based on beliefs that the past brought into being—there are no others.

If society does support potentially revolutionary science, then it cannot know what it will get. Why then even try to support it?

To evade this uncomfortable question, the scientific community has been able to convince itself and society at large that corollary benefits inevitably flow from advances in scientific understanding. Only minorities—so far—have suggested that this has not always been so; or, even if it has been so in the past, that there is no guarantee it will continue to be so in the future (let alone that the benefits will be in some proportion to the initial expenditure). So we spend billions of dollars on larger and larger atom smashers even while most scientists (other than high-energy physicists) believe that nothing of practical human use remains to be discovered along that direction.

If it is nevertheless decided to support pure science and to stimulate revolutionary discovery, it remains unclear how that might effectively be done. That the chief mechanism used by the National Science Foundation is clearly inappropriate does not necessarily mean that more appropriate mechanisms can be found. In point of fact, though, there are several more logically sound ones:

1. So-called starter grants, given to newly qualified scientists solely on the basis of their promise and not for projects that have to be specified beforehand in some detail. (A few such awards are in fact made by Research Corporation and by the Petroleum Research Fund of the American Chemical Society.)

2. "Established investigator" awards, similarly given to people rather than to projects, the people being selected on the basis of past accomplishment. (A few such awards are made by the Sloan Foundation and the National Institutes of Health.)

3. Prizes for past accomplishment, which in a sense is an established investigator award.

Naturally, such mechanisms have been used much more frequently by private groups than by governmental bureaucracies. The support of pure science—the search for knowledge—cannot come easily or naturally from government. If totalitarian, the government is tempted to draw its distinction between correct, acceptable knowledge and other knowledge that is to remain taboo. If democratic, the government feels obliged to account for its expenditures, and therefore to hold accountable those who do research under its support; and accountability eschews risk and seeks tangible results. But where concrete results are demanded, research becomes mundane, particularly when results are looked for within the usual lifetime of a research grant, typically no more than a couple of years.

Goal-oriented Research

The faith is widespread that support of pure science inevitably leads to startling discoveries that are not only palatable but even useful. Such faith is entirely compatible with the myth of the scientific method: if it requires only that the method be followed for true knowledge to be gained, then there is no evident reason why the method should not be applied to just those problems or phenomena whose understanding can be foreseen to be desirable and have useful application. Thus the myth of the scientific method leads to the misconception that science can give us the knowledge and applications that we want (or think we want). In support of this misconception such instances are typically cited as the development of atomic bombs or the putting of humans on the Moon: we often hear, "If we can put men on the Moon, surely we can cure cancer . . . or produce energy in a nonpolluting way . . . or eliminate poverty, et cetera, et cetera."

Such declamations fail to comprehend that the atomic bomb and the Moon flights are examples not of pure science but of applied science, of research into the known unknown (and into the rather well-known unknown to boot) to develop specific technology. The understanding—the pure science—already existed of how people could be lifted off Earth and guided precisely to the Moon; it was just a matter of constructing the mechanisms. It had also been understood how energy is produced when uranium atoms fission, and it was rather obvious what sort of devices could in principle harness the energy explosively or gradually (though it was not known whether such devices could actually work in the needed ways); enough was known to put thousands of people to useful work almost immediately on many details.

By contrast, the causes of cancer remain in all probability still largely in the realm of the unknown unknown—and they were even more so two decades ago when the government (incidentally, against the advice of the best-qualified scientists) allocated prodigious funds (of the order of a billion dollars per year) to finding a cure for cancer. Most disinterested observers now recognize that very little of value has resulted from this direct war on cancer; the most consequential advances in understanding have come instead from outside the war, from research into fundamental molecular biology, which attempts to understand how genes that control growth are switched on and off during development and in later life.

Goal-oriented research, applied science, makes sense only when there already exists the relevant basic knowledge to show that, and roughly how, the goal can be reached. Otherwise, resources can only be wasted and disappointment is certain. As Erwin Chargaff has pointed out, goal-oriented research without the requisite understanding is exemplified by alchemy, the attempt to transmute less-noble substances into gold. Emperors, kings, and other patrons kept many alchemists employed for centuries but got no gold in return.

Controlling Science

While it is quite appropriate to believe that applied science can be controlled, as already illustrated, we sometimes neglect the necessary qualification: "provided the necessary basic knowledge is already at hand." The really damaging instances, however, are those in which attempts are made to subject pure science, sheer knowledge and understanding, to ideological control.

Though we may know better in principle, in practice we keep succumbing to the misconception that we can control the outcome of science, forgetting to make the distinction between pure and applied science. Indeed, in common parlance even the distinction between science and technology, between knowledge and practice, is typically blurred, as when science is blamed for pollution or when even the editor of *Science* writes that "a dictatorship can put up with poor science if it can depend on espionage and imitation to produce modern goods." Now it is indeed possible to exert meaningful control over applied science (and over technology), but the outcome of pure science is *in principle* uncontrollable.

It bears repeating that our beliefs—the conventional wisdom, the scientific paradigm—are all the knowledge we have, and therefore we cannot allow for the fact that some of those long-cherished beliefs are themselves wrong. We cannot even predict which part of the conventional wisdom will fail first, let alone how. Science is knowledge, and we cannot know a thing before we know it; history teaches that shocks await us even as we least expect them. The greatest steps in the progress of pure science can only come serendipitously, and only as and when the state of assembly of the puzzle permits. We cannot control when they will come or what they will be.

Classic illustrations are widely known. Galileo, having submitted to the authority of the Church, which held the Earth to be central and stationary, could still mutter to himself, "And yet, it *does* move." When Nazis rewrote textbooks of physics, they still could not make physics work without relativity, even though the texts were written by a Nobelist.

When Lysenkoism was enforced in the Soviet Union, genes did not thereby disappear, only Soviet agriculture—applied genetics—suffered (and Soviet geneticists too). We do not learn from the past, though, and misguided attempts to control science continue, in democratic societies as in autocratic ones. Thus some current ideologues in the United States call for science to be different than it is: it should become "democratic," it should be "responsive to human needs," it should not be macho or power-seeking, and so forth. Such statements make no sense unless distinctions are made between frontier science and textbook science. The latter has survived because it reflects sufficiently faithfully the nature of the real world: the Ideal Gas Law is neither macho nor effeminate, neither democratic nor autocratic. Pleas for alternative science are simply absurd if they call for an alternative *textbook* science. Frontier science, on the other hand, is as human an activity as most any other, and ideological control of the players and of what they do and of what they say is not out of the question; it is just that history teaches that all such attempts stifle rather than improve science.

Attempts to control pure science are based implicitly on the fantasy that knowledge about nature can somehow be forced to be of a sort that we like. We are impelled toward that fantasy in some part by the misconceived fear that knowledge in and of itself determines action. That misconception is closely allied to what philosophers have long identified as the "naturalistic fallacy": that because a thing *is* so, therefore it *ought* to be so. Fallacy or not, we continually slip into it.

One root of the naturalistic fallacy lies in past centuries: since God created everything, and since by definition He knew what He was doing, therefore everything is the way it ought to be. Only the form, not the substance, of this misconception changed following Darwin: since evolution proceeds by adaptation and survival of the best fitted, and since evolution means progress, which we think is good, therefore things are the way they ought to be—and, for example, one should discourage the survival of the unfit.

Forgotten in this misconception is that humans and human societies are capable of caprice, charity, chivalry, generosity, mercy, wastefulness—all sorts of things that nature neither demands nor bars. Assume for the moment that there does exist "intelligence" as a significantly hereditary trait, that some individuals are fated purely by parentage to have less of it than others do: it is by no means a necessary consequence that the less intelligent be made inferiors in any respect *in social interaction,* let alone outcasts or slaves. That some identifiable groups of people characteristically carry certain "defective" genes again does not require that those groups should be prevented from reproducing, let alone that they ought to be massacred. There is no support in logic for the fear that *knowledge* will automatically lead to particular action.

The fear might then be rephrased thus: although logic does not entail it, human society is likely to go in that direction. But here history demonstrates the fear to be unfounded. Even if it could be shown quite beyond doubt that eugenics would lead to some generally desired end, that in itself would not ensure that human societies would practice eugenics. After all, it *is* beyond doubt that an increasing national debt and an unfavorable balance of trade must ultimately spell economic doom, yet countless societies have failed to balance their budgets or their trade. Just as with individuals, so too with societies: they have an almost unlimited capacity for rationalization, finding *purported* reasons— in science, in religion, or in whatever else might seem to carry the highest authority—to justify what they want to do for actually quite different motives. It is the exception rather than the rule in human affairs when people are actually guided in their actions by true understanding or knowledge.

We cannot escape the realities of nature. If intellectual facility is determined in part by heredity, then that will be so no matter whether we know it or not. But if it does happen to be so, that would not make it necessary that intellectual facility be made a *social criterion* of any sort.

Trans-science

The fallacy that what is so must and ought always to be so connects closely with the equally fallacious notion that science can and should decide questions of public policy. Adherents of the fallacy take the view that if science says that nuclear power is safe, then we should use nuclear power; if science says that the Star Wars defense system can work, then we should build it; and so on.

Questions thus put to science that in reality have no scientific answer have been called by Alvin M. Weinberg "trans-scientific." Any question of "ought" or "should" is by definition trans-scientific and has no answer within science, no matter how many technicalities happen to be entangled in it. To think otherwise is to commit the naturalistic fallacy.

But there are other sorts of trans-scientific questions where it is even easier to slip into the mistaken belief that a scientific answer exists or can be found. There are issues, for example, in which the inherent fallibility and limits of science need to be taken into account. How likely is it that an accident of a given type will occur at a nuclear plant? All estimates are based on belief, theory, guesswork, because not every possible cause can ever be *known* to have been identified. The only way to check such estimates of probability would be by trial: let 1,000,000 plants operate for 100 years, say, to check whether the probability of accident is indeed 0.0000001 per plant per year or whether it is higher or lower than that. But we cannot carry out such trials, and therefore our very estimates must remain uncertain. As Weinberg points out, many questions long debated in public are of this sort. Science can only place upper limits on how dangerous might be exposure to radiation or to any given substance: no matter how much experimentation is done, it can never be established beyond doubt that any specific thing *never* causes *any* damage.

How science can and cannot contribute to public policy is much too large an issue to tackle here, and it is not directly germane to the theme. What is relevant in both respects is the need to recognize that no bit of scientific knowledge by itself can force the adoption of any specific social policy, and that it is also a misconception to imagine that science can answer a question just because it happens to be posed in technical terms.

Technology

Misconceptions abound over the relation between science and technology as well as about each of those endeavors separately. An overwhelming majority of what is publicly discussed under the rubric of science actually has to do with technology: almost everything to do with medicine, for example, or almost anything having to do with pollution.

Perhaps the most prevalent fallacy about technology mislabels it as applied science. That seems plausible, of course: science is knowledge, and application of it could or should provide food, shelter, tools, and so forth. It just happens, though, that this has not usually been the way of it. True, scientific advances have sometimes sparked new technology: atomic bombs and nuclear power plants and transistors are the standard examples. But instances of technology leading to science are no less dramatic: electric batteries and magnetic fields, photography and cyclotrons, among others.

One reason for misconceptions about technology is that serious, systematic study of technology began so lately. The philosophy of science has been carried on for centuries, whereas the philosophy of technology has been a recognized specialty for barely a couple of decades. The history of technology, too, has been a distinct field for only a few decades, whereas the history of science can easily be traced into the previous century. That science and technology are quite different sorts of things, and that the relationship between them is anything but simple and one-way, has been made clear just within the last generation, most forcefully perhaps by Derek de Solla Price.

That technology is not just applied science follows obviously—once one thinks about it—from the historical certainty that significant techniques are ever so much older than anything one could call science: applications of fire to cooking, to extracting metal from ores, to metallurgy; levers, wheels, water levels—devices that made possible the wonders of megalithic construction and the American and Egyptian pyramids; pottery and glass; tools and weapons; tanning and weaving; et cetera, et cetera. Humanity created practical marvels long before the advent of what we call science.

The making of models to mimic the working of the solar system has been traced back many centuries, beyond times when they could have been based on scientific understanding of astronomy, and our clocks are descendants of those models and not timekeepers designed as such from first scientific principles. The crafts associated with such things, and with lens grinding and the like, were a significant factor in that ferment of the seventeenth century that led to modern science. Steam engines were developed through a series of inspired inventions, not through successive application of scientific insight; rather, these inventions led to scientific insight, to the science of thermodynamics. The lead-acid battery, ubiquitous in cars, planes, and boats, has been improved dramatically over the course of a century through improvements in material technology, not through advances in electrochemical science.

So science and technology have grown rather independently of one another—which does not mean, of course, in isolation from one another. They learn from and assist one another, but one ought to be clear that they are *in essence* different sorts of things—and much flows from the difference.

Notably, science is universal, whereas technology is particular. For instance, the behavior of gases is the same everywhere: the same equations govern the relations among pressure, volume, temperature, and amount of gas. By contrast, technologies can be vastly different: electricity can be made from nuclear power, or from falling water, or by harnessing the tides, or by burning any number of substances. Light can be made by burning things, or through chemical reactions, or from radioactivity, or from electrically powered devices, or by capturing and feeding fireflies.

Correspondingly, science has a continuity over time that technology has not. As science proceeds, much is continuous even before and after the so-called scientific revolutions: even as our mode of interpreting changes, perhaps quite drastically, our familiarity with natural phenomena nevertheless grows. But *technological* revolutions can make sharp breaks with the past, as quite different ways of doing something supplant one another—as the use of candles superseded oil lamps (which had succeeded open fires), to be in turn superseded by gaslight, and then by incandescent electric lights, which have been largely superseded by fluorescents. There may well be, as the jigsaw puzzle would indicate, a certain sequence of discovery that *has* to be taken in science; but in technology, that is much less the case, if it is indeed the case at all. If and when we make contact with extraterrestrials, we shall probably begin communication by means of the universality of mathematical and scientific laws. Extraterrestrial technology, by contrast, is likely to be entirely different from our own.

Science though universal is intangible: it is knowledge. Technology is inextricably bound up with tangible things (which accounts for its *not* being universal). Though advances in both science and technology flow from human creativity, they profit from different sorts of creations. Scientists all discover, or try to discover, the same phenomena and laws, whereas technological innovation can be unique. Thus creativity in technology is much more like creativity in art than in science: Wolfgang Amadeus Mozart, Salvador Dali, and Thomas Edison all created things that carry their individual stamp, whereas the Ideal Gas Law incorporates no clues whatsoever about the people who formulated it. Again, the laws of gases and elements and particles were bound to be discovered, perhaps even at about the same time as they indeed were, irrespective of the particular people who were helping to put the jigsaw puzzle together; but cars and telephones did not inevitably become ubiquitous, just as videophones and personal helicopters have not, despite their feasibility. As Erwin Chargaff has noted, it is not men who make science, it is the science that makes the men; whereas technology is intrinsically a human invention.

A significant consequence is that science cannot but be open and public, whereas technology can choose to keep its knowledge secret. Since there is only the one world to explore, credit and fame go to the explorer or scientist who discovers a given thing *first.* The best way to ensure that one's priority is established and recognized is to let as many people as possible know as quickly as possible what one has discovered; science and scientists are preoccupied with publication (moreover, with rapidity of publication). On the other hand, since technological products can be unique, secrets can be kept; so technologists say as little as possible about the crucial technical details of their ventures even after patents have been granted.

Science and technology have quite different criteria for whether or when something gets done. In science, putting the jigsaw puzzle together progresses best when each new piece is added just as soon as it is seen to be possible: if something is feasible, it will be done and ought to be done. In technology, on the other hand, the criterion is human benefit or utility: many feasible things are not done, and ought not to be.

Related to this difference is the fact that scientists and technologists look for approval in quite different directions. Scientists are first and foremost members of their scientific community; as they work and try to publish, at the back and front of their minds is always "What will they [meaning other scientists] think of this?" The reputation and worldly advancement of a scientist rests on the verdict of the scientific community. Technologists, by contrast, work to satisfy particular clients or employers, and they may not even have particularly strong associations with their peer technologists.

Science, as noted earlier, cannot be effectively controlled. It can be impeded or stopped, but it cannot be made to deliver particular knowledge that happens to be wanted. Technology, however, can be put under effective social control, and examples of that are legion. Thus domestic gas was made commercially from coal in Australia many decades ago, and liquid fuels from coal in South Africa, whereas such technology is still not deployed in the United States: economic and political criteria, not technical feasibility, determine technological development. (Of course, that technology can be controlled does not mean that all the effects of a new technology can be foreseen. Quite the opposite. It seems to be a law of technical innovation that every new technology brings some unforeseeable consequences with it.)

There are some typical practical repercussions of misconceptions about technology. Perhaps the most common follows from the mistaken belief that technology is applied science, for that implies that any advance in scientific knowledge could be harnessed to useful applications. As often remarked, this misconception is cultivated by scientists as much as by anyone, on the presumption that society will support science only if it believes that something of tangible use will result. But that presumption is wrong: we wanted to put people on the Moon out of adventure and competitiveness, not out of utility.

Another example of the confusion of technology with applied science is the fad that swept the United States, most noticeably during the last decade, for state governments to sponsor cooperative ventures between universities and industries in the belief that the latest science could thereby be quickly translated into new technology and corresponding economic benefit. This stampede, expensive in more ways than one, is based on considerable ignorance: of the distinction between frontier science and textbook science and the consequent fallacy of seeking quick application of the latest scientific work; of the related historical fact that, insofar as technology has drawn on science, it has drawn on *textbook* science; and of the difference in kind between science and technology, for it is not easy to envisage, let alone put into working order, a mutually beneficial and therefore viable cooperation when the interest of one partner is best served by complete openness while the interest of the other is best served by utter secrecy.

Social and Behavioral Science

It is quite generally felt that the social sciences somehow differ deeply and drastically from the natural sciences; in Britain, for example, "science" has been dropped from the erstwhile name of the Social (Science) Research Council. Yet what precisely is that difference? If anything, practitioners of the social sciences are much more explicitly scrupulous to follow the scientific method than are practitioners of the natural sciences. Why then are the latter so strongly and universally felt to be *more* scientific than the former?

If all the sciences were united and characterized by the scientific method, then sociology would differ from chemistry, say, only because its subject matter is different: more complex, it has been argued, or more inherently variable because it deals with living rather than with inanimate entities. Yet that response seems flimsy and unpersuasive because the perceived differences between all the natural sciences and all the social sciences are so great—they seem more like differences in kind than differences in degree. Sociology, psychology, and political science are very different not only from chemistry and physics but also from geology and biology, yet the latter deal with equally complex phenomena and with living entities. Even in those natural sciences that study very complex matters, a lot of things are rather definitively known, and a lot of things are rather well predictable even as they frequently run counter to commonsense expectation; whereas in the social sciences, little is known or predictable that is deeper than triviality or different from commonsense knowledge.

Once again, exposing the scientific method as myth opens the way to understanding. The social sciences can then differ from the natural sciences—just as the latter can differ significantly from one another—because the scientific method is *not* the criterion of whether a thing is science. Characteristically in the social sciences, by comparison with the natural sciences, little has yet trickled through the filters of consensus (see fig. 4); the action is still frontier or even pre-frontier, and much follows from that.

Here I shall argue that acquiring knowledge about the behavior of individual people or groups of people is *significantly* different from the acquisition of knowledge in the natural sciences. Reality therapy can often be conclusive in the natural sciences but only rarely in the social sciences; in consequence, a body of consensually agreed knowledge has accumulated in the former but not in the latter. Now it is the breadth and reliability of scientific knowledge that has gained for science its social prestige, so that it is inherently deceptive to label as science a field that proffers for most of its domain no useful knowledge map. Yet that is the case with the social sciences; and to support their claim to being scientific, social scientists adhere to and promulgate the myth of the scientific method—for their only possible claim to being scientific is that they practice that method.

Consensual Knowledge

In the natural sciences, the details of so many phenomena are now so well known that they can no longer be effectively denied even by determined minority groups: the courses of the stars and planets are accurately known, as are the characteristics of an uncountably large number of chemical reactions at work in our environment and in ourselves; we know that it rains only when there are clouds, and that the Sun's rays when focused can deliver intense energy, and so on and on. We know an enormous amount of the sort "If this happens, then that does too."

We prize the fact that, if we want to make an atom bomb or send someone to the Moon, all competent scientists agree on how to do it. (They will disagree over whether we ought to, of course, because that is not a scientific question but a trans-scientific one.) Several alternatives may be offered, perhaps by different individuals or groups, but they will largely agree with one another in assessing the relative strengths and weaknesses of the alternatives, and they will certainly agree with one another over how to go about resolving differences of opinion.

No such body of knowledge is available about human beings. If children grow up in traumatic situations, for example, we know only that some are likely to suffer from multiple personality disorder, that others will transcend their circumstances and become highly successful, and that the rest will cover a spectrum of life histories less emotionally disturbed than the former and less successful than the latter. Children of the same parents, even biological twins, develop differently, and we cannot tell parents, "If you do this, then that will follow," at least not with any assurance that our prescription will work a significant part of the time.

Or consider the community of social scientists asked how best to increase the efficacy of learning in elementary schools by, say, socially disadvantaged children. That there is no effective consensus has been proven clearly enough in the United States over the past several decades: the public desire has been there, the governmental desire has been there, there have been periodic infusions of appreciable sums of money, but no discernible progress has been made; indeed, it is periodically claimed that standards of literacy have continued to decline over this period in the face of a succession of supposedly ameliorative programs put in place by professionals in education, psychology, and sociology.

Or consider that there is bitter disagreement whether abuse of drugs would be more or less widespread if they were made legally available. Indeed, there is not even agreement with the benefit of hindsight whether the prohibition of alcohol in the 1920s in the United States was successful or counterproductive. Or, yet again, note that psychotherapists disagree among themselves over how to treat—and even diagnose—emotional illness.

We simply do not know very much, in anything that could be called a scientific way, about how or why people learn; nor do we know much in any scientific way about how people think, or about how they feel, or about how they are motivated, or about how those things interact to produce actual behavior. During some half a dozen years, while I was dean of a college of arts and sciences, I would visit periodically the various departmental faculties; and the social scientists I used to challenge: "Tell me something about human nature or human behavior that your discipline has discovered and that was not already known to, and written about by, dramatists, poets, and novelists." Very rarely would I be offered anything, and then it would be about something so limited and uninteresting as, say, the behavior of people in queues (in some given country at some given time, with no assurance that this could be related to what would happen elsewhere or at other times, what is more); or political scientists of the number-crunching sect would point to the accuracy with which their samples can predict the outcomes of elections—which may be a triumph of good sampling technique but is no demonstration of political or social insight. I was never offered anything approaching a biological theory of evolution or an atomic theory of matter—which was in a way surprising in view of the penchant sociologists in particular have to "formulate theories." Most striking, perhaps, was that the faculties could not even agree over what their disciplines might claim to have achieved.

It was difficult on those occasions not to be or appear offensive: inevitably it seems pejorative to suggest that any given academic field has not accumulated any substantial amount of reliable knowledge. At least it seems pejorative to suggest that *a science* has not done so. After all, philosophy, universally agreed not to be a science, is nevertheless a useful and respected discipline; but it does not deal in specific knowledge at all, in the sense in which the natural sciences do: progress in philosophy—insofar as one can or even wants to talk in such terms—is not a matter of knowing more, more precisely, about more and more things. Professors of literature may deal with a canon—a view of or familiarity with a body of literature and its creators—but once more that is not a body of reliable knowledge in the sense in which the natural sciences incorporate a body of reliable knowledge.

In philosophy as in literature there is no single paradigm; rather, there are disparate schools of thought over such matters as realism or the basis for interpreting texts. Thus, to say that a discipline lacks a single paradigm and has not accumulated a body of reliable knowledge is disparaging only if that discipline is a science, or supposes itself to be one, or has the ambition to be thought one.

So the point I make here is not that the social sciences are not interesting or valuable but that they are not sciences, because "science" *means* (admittedly, among and as well as other things) a body of agreed-upon and to-be-relied-upon knowledge. Such an accretion of comprehension may not of itself make something a science, but without it a subject cannot properly be called a science. The social sciences do not command such a coherent body of acknowledged fact.

That statement is usually held to be so unpalatable or derogatory that support for it had better be strong. Beyond the examples already given, consider how students are trained in the natural sciences and in the social sciences respectively. In all the natural sciences, university curricula are quite inflexible: largely series of required courses and prerequisites, the details of which vary little among different universities even in different nations. There is hardly a senior-level course that has no prerequisite, or, indeed, hardly a sophomore- or junior-level course without one or more prerequisites. Thus all prospective chemists must take some elementary mathematics and physics; all freshman chemistry courses cover much the same material; at least year-long courses in organic and in physical chemistry after that are universal, and in analytical and inorganic chemistry virtually universal; all curricula include substantial amounts of laboratory experience and almost always some undergraduate research experience. The number of available additional elective courses is larger than any one student can take, but it is often only about twice as large.

Almost nothing of this is the same in the social sciences (or in the humanities). Upper-level courses rarely have any prerequisites—and that evident circumstance, in and of itself, already demonstrates the lack of a coherent body of assented-to knowledge. There are far more elective courses available than any student can possibly take, by perhaps a factor of five or more. Curricula differ distinctively between universities, and they differ radically among different nations. There simply does not exist a set of facts, laws, and theories that every practitioner of any given social science must know, use, and build upon in order to be judged competent.

Another, related difference: the relationship between pure and applied is more problematic in the social sciences. For example, public administration, by intention primarily an applied field, insists on its separateness from political science, sociology, and psychology, which one might have imagined to be the repositories of the fundamental understanding on which administrative practice could draw. There is no comparable separation in the natural sciences: pure chemists may look down on applied chemists—and (with less conviction) vice versa—but neither group regards the other as not chemists or believes that separate disciplines and departments could or should exist. Again, a lack of shared interests among pure and applied was at the root of the schism that recently split the American Psychological Association in two. By contrast, the American Chemical Society has long served the most diverse groups within it—pure and applied, academic and industrial, as well as subdisciplines ranging from chemistry and the law through materials science to chemical education and the history of chemistry.

In point of fact, applied social science is much more like technology than it is like applied science. Like technology, its relationship to science—to underlying understanding—is patchy and variable: psychotherapists, for instance, are guided by long experience at least as much as by theoretical insight; so too with education or administration. For technology as for applied social science, there exists no obviously relevant corpus of scientific knowledge, sufficient to its purpose, upon which it can draw.

Consensual Understanding

Even more remarkable than the breadth of consensually agreed knowledge of phenomena in the natural sciences is the degree of unanimity that subsists over the explanation of those phenomena. Not only is there but a single set of chemical phenomena, there is also but a single theoretical approach used by all chemists everywhere for understanding the nature of chemical bonding, and the interactions among molecules, and every other major sort of phenomenon. The prevailing consensus is reflected in a single paradigm in each of the natural sciences—which paradigm, furthermore, overlaps and is compatible with those of the other natural sciences. There is very considerable agreement in each field over what the present state of the art is, about what it means, and over what the most interesting current challenges are—in other words, over what parts of the jigsaw puzzle should be worked on next.

In stark contrast, the social sciences—collectively or individually—possess no governing paradigm. In each field there exist distinct schools with different beliefs about quite basic issues; and though there may be agreement over the existence of quite a range of phenomena, nevertheless one finds different and competing theoretical approaches to the understanding of those phenomena. Thus in psychology the schools of thought range from essentially mystical at the one extreme through psychoanalysis and the like all the way to the other, mechanistic extreme of Pavlovian behaviorism. In sociology the differences cannot even be described on a single spectrum—structuralist and relativist approaches, for example, refer to viewpoints based on disparate criteria of what is important—and it is more difficult to attach meaningful labels to sociologists than to psychologists. In political science, political philosophy and quantitatively oriented political science are worlds apart.

These various schools of thought differ from one another much more profoundly than do the subspecies within a natural science. In chemistry, say, the theoreticians and the experimentalists may dispute the primacy of evidence or theory, but they largely agree about what chemistry as a whole already knows, about how chemistry relates to physics, biology, geology, and astronomy, and about which bits of new chemistry would be most valuable. On the other hand, in psychology, say, the various schools of thought diverge even over what the subject of study is: whether it is some pattern of physiologically determined responses or an entity capable of exercising choice over its responses.

Controversies in the natural sciences have to do with points of detail (except, of course, during the rare revolutionary epochs)—matters of passionate interest perhaps to the few who are engaged in the actual investigations but of no concern to other scientists who know that the dispute will, in its own good time, be resolved within the paradigm. Disagreements in the natural sciences are often settled quickly, as all sides agree on fundamentals and on approaches and on what the evidence is. Disagreements in the social sciences, however, continue over many decades because they concern fundamentals rather than details; thus relativist and realist schools of thought have persisted and can persist in the same way as, in politics, Republican and Democratic parties could persist: these are thoroughgoing divisions (albeit not always easily defined) that supersede and outlast agreement or disagreement over matters of detail.

Coherence in the natural sciences exists not only in range, in that paradigms overlap and are mutually compatible across the whole of natural science; there is also coherence in any given field at the various levels of detail. There is general agreement, in other words, about how intricate details relate to the larger picture. No matter how narrow the specialization of a scientist may be, everyone knows in what part of the puzzle the work belongs and what the relation is between the shapes and colors and connections there and the corresponding characteristics of the larger piece of the puzzle in which that particular study is embedded; the work is significantly influenced by the prevailing paradigm and it can also significantly influence that paradigm: a measurement of the speed of light, on the face of it uninteresting to all but a handful of physicists, led to the theory of relativity, which dramatically changed the overarching theoretical scientific paradigm; or preoccupation with the spectrum of colors radiated by a hot body, again of little interest to most scientists, led to quantum theory, which influenced much of physics and chemistry very immediately. So investigations in science that sound trivial to outsiders can and do have meaning for science as a whole, because they cohere with it. Not so in the social sciences, where the adherents of a given school of thought can and do ignore much that is produced by the adherents of other schools; for example, what is done by proponents of the "public choice" school is not necessarily seen as relevant by Marxist economists, and vice versa. Consequently, it is difficult to show that or how trivial-sounding investigations in social science bear on our understanding of human behavior as a whole.

Reality Therapy

The manner in which detail relates to the whole bears on the much-argued issue of objectivity. In the natural sciences, almost all work now proceeds at so deep a level of minutiae that it has no directly evident relation to questions that really matter to human beings as such: a chemist's drive to synthesize some new substance, for example, seems hardly relevant to the possible survival of human souls after death, or to any other religious or social or political concerns. Consequently, most of the work done in science can be less hampered by wishful thinking or fanatical partisanship (although it must be admitted that scientists can become fanatically partisan about their pet theories!): it may not be that scientists can be entirely objective, but at least they can be largely disinterested in much of their work. They suffer conflicts of interest between acquiring knowledge and making a career, but not that the knowledge they gain might contradict their religious or political or moral beliefs.

Because science coheres, however, the *results* of studies of minutiae are able to influence the larger paradigms that do have a bearing on questions of genuinely human concern. So Darwin, for instance, had no call to foresee that his naturalist studies would have any necessary or even conceivable relevance to religious matters; yet his conclusions, constrained by much intricate detail, turned out to force changes in the prevailing paradigm, which in turn was taken to have relevance to religion. A *direct* attack by science on religion at the level of evolutionary paradigm could hardly have succeeded; but the issue is moot, because no such attack would even have been mounted.

In the social sciences, by contrast, it is very difficult to find even trivial things to study that do not have some immediately evident bearing on political, social, or religious doctrine; and so from the very beginning of studies of apparent minutiae, opposing schools of thought will disagree. The natural sciences approach objectivity because answers about minutiae, which are of no human concern, turn out to coerce paradigmatic views, irrespective of whether those are palatable or not. In the social sciences, preexisting and contrary paradigmatic views can vitiate the possibility of meaningful agreement even at the level of minutiae.

Social scientists themselves acknowledge something of this state of affairs: they describe their fields as multi-paradigmatic (or sometimes as pre-paradigmatic). What they usually fail to acknowledge, however, is that everything we admire and prize about the natural sciences—about science, in other words—stems from their *not* being multi-paradigmatic, from their accumulation of consensual, reliable, useful knowledge under a single paradigm (or, as already noted, under mutually compatible and overlapping paradigms).

The Knowledge Filter

The filter and puzzle model describes how the practices and institutions that have evolved in science sift out bias, error, and fraud under the scrutiny of the scientific community and the control of the prevailing paradigm. Now the appurtenances of those mechanisms are to be found in all the academic disciplines including the social sciences: training in undergraduate and graduate programs, reviewing of grant proposals, refereeing of manuscripts for publication, and so forth. The difference is that these mechanisms have not brought about in the social sciences the range and depth of overarching consensus that prevails in biology, chemistry, and all the other natural sciences. Why not?

In the natural sciences, reality therapy and the validity of textbook science are inescapably effective in nudging people toward consensus. When dealing with nature, one cannot easily or long evade reality: eggs will not cook unheated, for example, nor will bodies move by themselves. Scientists find it natural that certain knowledge about the world is accessible and that they must bow to it. By contrast, reality therapy is weak and ineffective in the social sciences. Able, accomplished, erudite, intelligent, perceptive individuals have disagreed for millennia, and continue to disagree, over fundamental questions in social science: Are there any inherent intellectual or emotional differences between men and women? What types of inherent tendencies can be eliminated or superseded by environment and upbringing? Does punishment deter crime? Are totalitarian regimes more efficient than democracies? more stable?

Different experiences of reality therapy may explain a stereotypical difference in approach to controversies, as between scientists and social scientists. Scientists believe that arguments can be settled if the facts are known, whereas social scientists are clear that disputes get settled by exercise of power, or by negotiation, almost irrespective of what the facts might be.

Similarly, scientists in practice label a thing pseudoscientific if the *claimed knowledge* is sufficiently implausible, even when no error in method is demonstrable (see chap. 3); scientists also accept new findings according to their plausibility, believing that if the results are OK then the methodology must have been appropriate. By contrast, social scientists tend to accept any finding, no matter how implausible, if the methodology used was sound. In their anxiousness to establish sociology as scientific, in point of fact, sociologists make a fetish of how their discipline differs from common sense (a strategy that naturally brings with it many problems).

In the natural sciences, disagreements can be quite separate from judgments of competence. But the very judgment of competence in the social sciences can be problematic as it rarely is in the natural sciences. In psychology, behaviorists and psychoanalysts and other schools as well disagree over what constitutes adequate training: does, for instance, a therapist need to have a doctorate, and if so should it be a doctorate in medicine or in psychology? The formal evaluations that social scientists make of one another speak typically of methodological facility and of numbers of publications, not of bits of new knowledge gained, whereas in the natural sciences such evaluations mention first and chiefly what discoveries stand to the individual's credit. Again, the refereeing of articles in physics has been found to be more consistent, based on an apparently more objective consensus, than in psychology.

The Scientific Method

If science is marked by consensus, then social science is not science. But this conclusion remains unpalatable to most social scientists because "scientific" in our society carries such high prestige and "unscientific" is so pejorative. It is only natural, then, that social science cannot bring itself to recognize that what truly does characterize science is consensus induced by reality therapy and attested by an established paradigm: to recognize that would be to admit that the social sciences are not science—not that they are simply young sciences, or sciences grappling with greater complexity, or a multitude of little sciences waiting to coalesce, but that they are simply not at the moment science.

Now social science has had at hand an effective and plausible way to evade this reality: it has clung to the myth of the scientific method. Under that criterion, if under no other, social science can call itself scientific: if all it takes to be scientific is that one frame hypotheses carefully and test them as best one can, and if being quantitative ensures that the tests are well done, then clearly sociology, political science, and the other social sciences are just as scientific as biology or chemistry or any of the other natural sciences. Indeed, social scientists are much more consciously scrupulous to follow the scientific method than are scientists themselves; they do so quite explicitly, in fact, whereas practitioners of the natural sciences simply *assume* that they are following the method, irrespective of what their actual practice happens to be (because they know themselves to be scientists and they have heard that scientists use the scientific method, whatever that is).

That the social sciences do define "scientific" as "using the scientific method" is illustrated by the sections on scientific method in introductory texts in psychology or sociology. For example: "The common element [in the sciences] is the method. . . . [What] makes a study scientific . . . is *how* one studies whatever he studies"; "Sociology is a science to the extent that the sociologist utilizes the appropriate method"; ". . . the method used . . . determines whether any study . . . is scientific"; "Sociology is considered a science because it uses the scientific method"; ". . . sociology is relatively less advanced . . . than most of the natural sciences . . . [because] the scientific method has been applied to the study of social behavior only in recent times, whereas the scientific method has been applied to the natural world for centuries . . . and the study of human behavior presents many problems that natural scientists do not have to confront."

That the natural sciences do *not* define themselves in those terms but rather on the basis of *the specific knowledge they have gained* is equally illustrated by the *absence* from introductory texts in chemistry or the other sciences of any substantial discussion of the scientific method: at most there are a few paragraphs, in the somewhat half-hearted or even apologetic tone that also characterizes their references to the history of the subject or its applications to societal benefit—these are seen by scientists as "frills" about whose value they are ambivalent. What natural scientists are not at all ambivalent about is the actual phenomena they know and their ability to explain those phenomena. Social scientists, on the opposite hand, are ambivalent (or multivalent) about their phenomena and the explanation of them, but they are not at all ambivalent about what the scientific method is, that they employ it, and that they are therefore professional scientists. They so describe themselves, incidentally, and as "professional," very much more frequently than natural scientists do.

Whether a discipline can be deliberately made scientific by conscious deployment of the scientific method, and whether that enhances the search for knowledge in that field, was put to the test in archaeology. Beginning in the 1960s, archaeologists looked to the renaissance in the philosophy of science as offering their discipline practical guidelines for making progress by becoming self-consciously scientific. Some twenty years later, the consensus was that this hope had been illusory.

Not Science, yet Worthy

The questions of interest in social science are not answerable in the way questions in the natural sciences are. Are there inherent intellectual or emotional differences between men and women? What inherent tendencies can be modified by nurture? Does punishment deter crime? Are totalitarian regimes more efficient—or more stable—than democracies?

Careful attention to such questions reveals that they are not answerable as posed. Either they use categories that have been postulated by human beings and are not specified by nature or they hinge on judgments of value, not on matters of fact. Thus, "emotional or intellectual differences between men and women": we all know well enough what we mean by that, but when it comes to finding proof that any two individuals do or do not differ in those respects, the problem begins to evidence itself. To compare, we must measure, but here we do not know how or what to measure. As to environment and upbringing, nurture or nature, the debate has a long enough history and lack of resolution to demonstrate its intractability. The questions that are of inherent interest in the social sciences are, in the terms introduced earlier, trans-scientific: they can be framed as though technical answers exist when, in point of fact, they do not. Perhaps it is not even possible to find a question in social science that is at the same time not trans-scientific and not trivial.

In what sense, then, are the social sciences actually science? They have no unifying paradigm or the intellectual consensus that goes with it. They have not produced distinctive and reliable knowledge that is respected or valued by human society as a whole. Yet those are the very qualities for which the natural sciences are noted and respected; they are the qualities that we associate with something being scientific—that is, authoritatively trustworthy. The social sciences are simply not, in the accepted meaning of the term, scientific. And that conclusion has been reached by at least a few practicing social scientists—for example, Ernest Gellner, who wrote: "we shall know that the social sciences have become scientific, when their practitioners no longer claim that they have at long last stolen the fire, but when others try to steal it from them; when the philosophy of social science becomes a search for an ex-post explanation of a cognitive scientific miracle, rather than for a recipe or promise for bringing it about."

Indeed, much as social scientists press to be regarded as scientists, they do not really so regard themselves. At an early meeting of the Society for Social Studies of Science, a speaker (a sociologist) asked those in the audience to raise their hands who were or had been scientists. Up went the arms only of those few of us who had practiced chemistry, physics, or engineering: the political scientists sat still, as did the sociologists, historians, and others.

Now it is understandable, given the prestige science enjoys, that others might wish to share in that; and promulgating the myth that anything scientifically methodical is science would appear to be a way of doing that. Yet it is also curiously nearsighted, for the modern understanding of scientific activity hinges directly on understanding the individual and collective behavior of human beings. After all, it is human beings who have evolved the practices and institutions that govern science; psychological and sociological insight are inescapably needed if the nature and ramifications of scientific activity are to be understood. Virtually all the characteristics of frontier science reflect human psychology and human interaction, and the transmuting of this most tentative science into reliable textbook knowledge demonstrates how the working of social institutions has intellectual corollaries. To understand science we need psychology and sociology as well as history and phi-

losophy—albeit we need them first and foremost in their interdisciplinary guise within science and technology studies, not in their common
disciplinary attire.

The social sciences are no less potentially valuable than the humanities or the natural sciences, and they are needed for the proper comprehension of scientific activity. Yet they are not sciences themselves,
for they do not exemplify the needed quality of consensuality. Effective
consensus springs only from successful reality therapy: there is no other
way to achieve a *voluntary* consensus of *rational* opinion. But reality
therapy can be successful only if a given cause always has the same
effect, given sufficiently similar circumstances. As social science deals
with human beings, it is not clear that "sufficiently similar circumstances" are achievable or that a given cause—even if it could be disen-
tangled from other contributing ones—would always have the same
effect: after all, human beings might indeed in some measure possess
the quality that we call "free will."

That something is not a science does not make it useless. Any insights
that psychology, sociology, and the like can offer us will be gratefully
received, whether or not they claim to be scientific insights.

7

In Praise of Science

That science is inescapably a human activity does not mean that it is *only* or *just* a human activity, *essentially* similar to all other human activities. Some modern pundits of science nevertheless seem intent on denying science any special status by emphasizing only how similar the behavior of scientists is to the behavior of other people: "Science is business in disguise. Scientists sell knowledge-products as commodities in a competitive market; their livelihood depends on it. A variety of buyers (industry, the state, venture capital) choose to buy science. . . . science is a commodity not unlike deodorant or mouthwash."

By substituting "art" and "artists" for "science" and "scientists" in the above, the sense and validity remain much the same; by substituting "education," again the passage conveys just as much. All human activities through which people also earn a living share these characteristics, and it tells us nothing about science, specifically, to point them out. For some people, the way they earn their living is only that; for others, whether or not they earn a living is immaterial so long as they are doing what they want (or must); and the continuum between those two extremes is heavily populated by people who have a mixture of motives for the manner in which they spend the time for which they are paid. Unskilled laborers, shopkeepers, sociologists, doctors, and ministers all get money for what they do; and all of them advertise the value of what they do, partly out of conviction and partly in the hope of higher remuneration. But that does not make all those enterprises in any meaningful way "business in disguise." If one seeks to explicate what science is, why it is not the same thing as art or education or trade, then the relevant points are those that distinguish science from the other pursuits. Pants and jackets are both clothes, but if one describes only what is similar to both then one neglects genuinely important differences; one might even say that almost the least interesting or significant thing about jackets and pants is that both are items of clothing.

To think of science as a business that advertises its wares and sells a commodity is no less misleading, no more illuminating, than to think of it as the dispassionate pastime of practitioners of the scientific method: both contain little grains of truth, but through being unqualified they hide the real or whole truth. One can understand how science works and interacts only by recognizing that scientists are caught in a conflict of interest between such extremes, between creating knowledge and earning a living; and that scientific institutions are similarly constrained in choosing what problems to attack by the high cost of modern research.

P. G. Abir-Am, in "Toward an Ethnographic History of Molecular Biology," interprets the fiftieth anniversary of the first X-ray photograph of a protein. She "deconstructs two key 'documents,' the invitation and the program, as textual strategies designed to mobilize the 'tribal assembly' for a succession rite in which three successor heroes divide among themselves the legacy of a venerated 'ancestor.' . . . the three successor-heroes . . . shared their recollections of their respective 'initiation rites' . . . while emerging as legitimate spokespersons-for-a-field's history." That scientists in groups show similar characteristics as other people in groups tells us nothing about why science has the special status and importance that it does. It is useful in *social studies*, of course, to adduce examples of behavior from a variety of fields; but it does not help the study of *science* to point to similarities rather than distinctions. Here is another illustration of the need for genuinely interdisciplinary studies of science, technology, and society: sociology focuses on the generalities of social behavior, and the sociology of science (as in the pieces cited above) draws on science for illustrations of that behavior, whereas STS is concerned directly and primarily with the understanding of science in itself, bending to that task whatever insights any of the scholarly disciplines can lend.

The insights of practicing scientists and engineers are needed as much as those of historians, philosophers, and others if the nature of science is to be comprehended and the comprehension fruitfully acted upon. Students of science must have a feel for the sort of thing science is, and they must have respect for it—students of any subject must have some feel and respect for it if they are to generate authentic insights. All too often, however, as in the pieces quoted above, one senses a hidden agenda of decrying or deploring, of dragging science off its supposedly undeserved pedestal. In railing against scientism, some pundits find themselves casting stones instead toward science itself.

Vive la Différence!

Science is uniquely distinguished from other human practices: it is the only activity in which the constraints of reality have brought to the quest for deep answers an effective consensus across all the variations that in other respects divide the human species. The accepted findings of science are the same in all countries, in all languages and for people of all ages and religions and genders. Only in science has such consensus been achieved *through the voluntary assent of all concerned.* In other disciplines, schools of thought continually dispute one another with varying degrees of intensity. In everyday political or religious affairs, consensus has not been achievable even through warfare and torture.

Familiarity does breed contempt. We are so accustomed to the international and universal character of science that we easily fail to remember what a marvel that is. In all the other branches of knowledge intellectual dissension reigns even among people who in other respects share a great deal. In most every department of English, for instance, colleagues are at one anothers' throats over the criteria for interpreting works of literature, over the respective value of different items, over judgments of competence: over just about every important thing having to do with their expertise and their profession. Musicians and consumers of music are at loggerheads over the merits of atonal as compared with classical music; religion has served to divide human beings into a large number of hostile camps; political allegiances cause bad feeling and worse; and so on. Human beings disagree about everything under the sun *except science* (and specifically, of course, *textbook* science, for frontier science is not the place where consensus reigns).

As John Ziman has shown, the innumerable little details of everyday scientific practice, as well as the larger picture, can be understood to quite a decent depth of sophistication by simply regarding science as the attempt to gain a *rational* (which connotes voluntary) consensus over the widest possible field. That the attempt to attain consensus has worked so well in science but not in other spheres is owing to the peculiar effectiveness of reality therapy when one studies nature. If what one seeks to explain or understand or control has to do with rain, for instance, then it is simply no good trying to maintain that it is raining when it is not, or that five inches have fallen when containers have gathered only one inch of water: one's assertions are too easily contradicted by plain facts that are evident to all other people. One can fool all the people some of the time, and some of the people all the time, but one cannot fool all the people all the time *when the evidence is as clear as it can be in natural science.* Nothing, by contrast, can force one person to agree with another about which approach to literary criticism is the best, right, or most fruitful; and we simply do not know what makes some children grow up curious and others uninterested; and we can and do argue and disagree over such matters without end.

Science Is Not Scientism

The marvelous successes of science led during the nineteenth century to the belief that other fields could be as successful if one only applied to them whatever the secret of science's success is. Hence the view—usually called, pejoratively, "scientism"—that science and only science is capable of generating true knowledge, and moreover that science can generate any true knowledge that we might wish to have. Of course, the supposed secret of science's success was the scientific method.

Now that science has turned out not to be the panacea for all the material problems and spiritual questions that humankind chews at, a natural reaction is naturally swinging far toward the opposite extreme. Science's knowledge differs from other knowledge only in degree, it is said, if at all: all knowledge is a human construction, not a reflection of the outside world; science is not always objective, therefore scientists are not more objective than anyone else; and so on.

But such a petulant reaction is quite unwarranted. That science does not have all the answers does not mean that it has no answers. That science now has inadequate answers in some areas does not mean that the answers will not become adequate in the future; in fact, history teaches that science's answers become better and better as time goes by. That science is fallible does not mean that science is *entirely* fallible or that it is *as* fallible as such other modes of human knowledge and belief as folklore, religion, political ideology, or social science. That science has no answers in some matters—such as the value of human life or the purpose of living—does not mean that it has no answers in other areas—those areas that are within its purview, matters of forces and substances and natural phenomena. And that science has no direct answers on matters of human purpose does not mean that its answers on other matters have no bearing on how, and how well, we are able to think about human purpose, free will, and other such things.

The fact is that science has created a remarkably wide-ranging and coherent understanding of a great deal. We can tell accurate stories—accurate in that they are good maps—in considerable detail about happenings on scales smaller than a hundred-millionth of a centimeter and during times shorter than a million-billionth of a second, all the way up to dimensions larger than billions of light years and time spans of billions of years. Through characterization of the hundred or so elements and their properties, we understand the existence of untold millions of different substances, and much about their individual as well as generic properties; and we even understand—already, after barely a few hundred years of real scientific activity—quite a lot about the phenomena that distinguish what we call living matter from inanimate matter.

The range of contemporary scientific knowledge is staggeringly great, and the reliability of that knowledge is notably robust. Most often, when this is acknowledged it is connected with technology, with the power that science has given humanity over the environment and over other living species. Not so often emphasized is the degree to which scientific knowledge has enabled humankind to understand the world and to admire and appreciate it in a relatively realistic and unfearful manner. Before science, the events of the world could only be explained as the doings of more or less anthropomorphic (or zoomorphic) beings, spirits both beneficial and harmful, gods and devils, entities about whom one could guess but about whom one could never be quite sure. So anxiety could never be far distant, unless it was kept at bay by unquestioning (which is to say fanatical) beliefs about the characters of those gods and devils—beliefs that clashed always with the equally fanatical beliefs of other human groups and therefore led to wars and torture and to the attitude that such means could be justified as necessary. Nowadays for many people the understanding gained through science has banished much of that pervasive anxiety, and for many groups the understanding gained through science has led to mitigation of the clash of religious beliefs. Science has been at least as important in human affairs through its influence on our thoughts and beliefs as through any material consequences; the most important benefits of science are cultural ones, not material ones.

Scientific Literacy and Education in Science

Scientific literacy—if one must use the term—ought to mean understanding something of that, being able to see science as a strand in the intellectual and religious history of humanity. Scientific literacy is part of being historically literate, of being culturally literate—in short, of being literate. I have argued that one can be literate about science without knowing much of the content of science, by comprehending that in working the jigsaw puzzle, progress is constrained by the filter of scientific consensus, which works better as more use is made of reality therapy. That view enables one to think intelligently about the impact of science in human affairs, to arrive at sensible answers to significant questions, as I now seek to illustrate.

How authoritative or reliable is science? Answer: It depends whether we are talking about textbook or about frontier science. Ask the scientific community. If there is consensus, *and if the knowledge is maturely seasoned and explicated in textbooks,* then you can safely give odds of better than 10 to 1 that it is trustworthy. If it is newly minted knowledge, *even if the experts are all or almost all agreed,* you should not give nearly such good odds on it. And if there is no consensus, you had better act on the basis that no one really knows.

Being scientifically literate should mean being able to answer in that sort of fashion. And it should mean understanding that, by contrast, one would not bet, at any odds, on predictions made even by a consensus of social scientists about something in a textbook of social science. And that we should not expect literary critics or ministers of religion to know anything useful about the stars, the Earth, the weather, or biology; nor attend to such other nonscientists as astrologers, psychics, and the like about matters having to do with nature.

The scientifically literate should understand that when we want to know why we are alive, whether that has some higher purpose; or how we should live, to feel progressively more satisfaction; or whether a God governs the universe . . . that it is then no use—or worse—to ask scientists or science. Then is the time to commune with social scientists, ministers, historians, philosophers, and the like.

What is the social value of science? Why should we support it with taxes? Answer: It can keep people honest. Emperors and popes used to insist that people subscribe to lies about the Earth, about the relationships among different sorts of people, and about a lot of other things. They cannot lie to that extent anymore. Science can put and keep politicians and prophets in their proper place, at least over some things.

Science can be endlessly instructive. One can be quite literate without knowing very much of the content of science, but one still remains sorely unlearned and ignorant who does not know the outlines of what science has to say about the histories of the universe, the Earth, and life on Earth. Ignorance is attendant to superstition, and those who have no understanding of the world "may invent mysteries, new gods, much as people did around lightning and eclipses, around St. Elmo's fire, and volcanic sulfur emissions a long time ago."

Studying science is excellent training for the mind, much better than the classically prescribed study of Latin. When you study science in the right way, you learn about reality therapy; and that is worth applying to other things than science. Science can teach that some things are quite definitely wrong; that knowledge is a much better guide than ignorance; and it can teach humility in posing endless questions to which we have no good answers.

Science can benefit our health, our pocketbooks, and our ability to have interesting, pleasant experiences. Knowledge that was newly discovered ten, fifty, or a hundred years ago has now become solid enough that technologies use it. If we want more control over our environment a hundred years from now, then we should continue doing science.

Scientific research is an investment in the future; trying to make it pay off quickly is as counterproductive as is, in the economic sphere, skimming wealth from corporations through leveraged buy-outs instead of investing for the long haul. Science is part of humanity's cultural heritage. Being educated in science is as important as being educated in philosophy, or psychology, or foreign languages because without it one is ignorant, a primitive savage rather than a civilized human being. And to be scientifically literate is to understand that.

The Scientific Method as an Ideal

That scientists in practice do not actually use the scientific method, and that the scientific method cannot adequately explain the successes of science, does not mean that the method is not worth talking about, that it is not worth holding as an ideal. After all, that we are all bound to sin does not entail that we should not strive to avoid sinning; differences of degree can still be differences that really matter.

Dispassionate, objective, systematic pursuit of knowledge is profoundly unnatural. Considering what the course of biological evolution is likely to have been provides ample clues as to why this might be so. For sheer physical survival, knowledge provided by immediate experience needs to have a large impact, so that imminent dangers can be avoided through quick action or sudden opportunities grasped—"instinctively," as we say. The ability to draw conclusions from a whole pattern of circumstances lends speed of decision by contrast with analytic consideration of the combined effects of many different factors acting together. The drives of strong passions will have ensured the reproduction of genes or patterns of genes for decisive action in preference to dispositions that might favor disinterested judiciousness and delay in decision making and in acting.

But one does not need to turn to that sort of speculation, for the evidence itself is plain and overwhelming enough that human thought is indeed by nature impressionistic, illogical, and prone to serious error, that it proceeds in some holistic or pattern-searching manner rather than along analytic and cause-considering paths. Human beings are particularly bad at systematic collecting and recollecting of information; at any processing that involves probability or statistics; and at modifying belief under the influence of reality.

As we try to amass information, anything that happens to us individually or directly carries far more weight with us than it logically should. Consider, for instance, how one's opinion about a given model of automobile is influenced by the stories of friends who have owned one, in comparison with the copious statistics offered in such surveys as in *Consumer Reports:* whereas the numbers should in logic count for much more, we tend to pay more attention to the graphic anecdotes of those whom we know personally. Or consider how we are struck by "the meaning" of a coincidence, when by definition and by logic a coincidence has no meaning at all because it results from the random workings of chance.

So far as our ignorance of probability is concerned, one of the standard illustrations is the birthday problem: how many people must there be in a room so that there is a 50:50 chance that two of them have the same birthday (day and month)? Few people can calculate the correct number, which is 23; and quite a lot of people refuse to accept that answer even when the reasoning is explained. Human intuition simply does not instinctively draw upon the laws of probability, yet those laws accurately reflect an important part of the reality of the outside world.

Scholars too typically lapse into illogicality; and scholars, like other human beings, find it difficult to modify beliefs once they have been acquired. We have surely all observed how even some of our friends, let alone other people, can remain impervious to the most cogent arguments supported by the strongest possible evidence, how they manage to sustain unwarranted, illogical beliefs—often enough beliefs that mutually contradict one another. (What we observe less frequently, of course, and with more difficulty, is that we are just like our friends in that respect, and like all other people too.) It is part of the human condition that we take personally whatever happens to or around us, instead of comparing the experience with a control (the probability that such a thing might happen by chance). It is part of being human that we explain away occurrences that do not fit in with our preexisting belief rather than use the contradictory data to modify our belief.

But even as those things are part of the human condition, human beings do differ in the *degree* to which they have *learned* to be otherwise; and one of the values of science (perhaps even the greatest value) is that it has brought humankind some glimpses of what successful, realistic ratiocination is. For many centuries certainly, in all likelihood for many thousands or tens of thousands of years, human beings literally did not know what to believe or why. Children believed their parents, by and large; people believed their leaders, by and large; the laity believed religious leaders and religious texts, by and large. But every now and again there would come clues that beliefs could actually be tested, and conflict between traditional believers and would-be empiricists doubtless dates far back into prehistory. In historical times, one sees in the events of Western civilization the splintering and recombination and struggle of religion to strike an effective balance between revealed truth, on the one hand, and experienced reality, on the other. Some centuries ago, empiricism turned out to work, provided people banded together to keep one another honest, to prevent prior beliefs and wishful thinking from vitiating the learning of new things; and soon people began to recognize how well the empirical method could work in the study of nature. Humanity has hardly looked back since then.

The various attempts to formulate the scientific method can properly be seen as attempts to find ways in which people, by nature illogical and immovable in their beliefs, might learn to be less illogical and to fool themselves less. That false belief can and should be rejected under reality therapy is anything but a trivial insight, though it seems so obvious to us now: it seems obvious largely because the progress of natural science, over several centuries, has convinced us of the power of that approach. There are large and small groups of people to this day that do not accept the priority of evidence and experience over revealed truth.

So science has shown humankind how to learn; and the scientific method specifies some rules that, if followed, permit one to learn. Thus the rule that we must make our hypotheses—our beliefs—explicit before they are tested is a necessary one: it is more difficult to slip out of a well-defined position than out of a poorly enunciated one. (That too is why we sometimes gain from individual psychotherapy: by enunciating our beliefs freely we can—especially with the help of a disinterested observer—come to recognize that we believe things that we didn't think we did, or that contradict some other things that we believe, and that cause us to behave in ways that puzzle us.)

The scientific method represents an ideal eminently worth striving for, not only in science but in all fields. To the degree that we manage to deploy it, consensus can grow and conflict decrease. But we ought to be clear too that no individual can follow the method unaided, because it is so inhuman an approach, foreign to our way of thinking. We need the help of peers—teachers, referees, reviewers, colleagues, competitors—to enable us to see where and when we are insufficiently explicit in laying bare the presumptions underlying our hypotheses, how the tests we design are not sufficiently stringent, what the loopholes are in our reasoning. What makes science, the study of nature, so special is that overlying all our wishes and strivings to learn, to be realistic, to be sensible, stands the help of nature, whom we cannot force to behave in any other way than what comes naturally.

Our attempts to apply the scientific method to human affairs have not been successful because it is so much more difficult there to demonstrate that a hypothesis indeed needs to be abandoned: the tests are difficult to design and the multiplicity of contributing causes allows plenty of room to explain things away rather than explain them. The nineteenth-century view that science is what it is because of the method made it seem that all features of society could be made objective and rational since the method is inherently so simple. But now we understand that the scientific method is an unattainable ideal; that we approach it to the degree that social institutions hold it as an ideal; that it works only according to the degree to which individuals and groups are willing and able to submit to reality therapy. Thus nineteenth-century optimism can now be seen to have been at the very least premature. But that reality therapy has not yet been effective in psychology or sociology does not mean that it will never come to be effective, let alone that we should cease from the attempt to make it so.

Even if social science never becomes actual science, natural science will remain a most remarkable human achievement, the greatest of human intellectual successes. Its content, and the manner in which it was acquired, remain useful and instructive. Its universal character remains the only thing that has allowed (some) human beings to communicate with one another despite all barriers.

To say that the scientific method is a myth is to say that it is not literally true, which is not the same as calling it worthless: "Myth expresses, entrances, and codifies belief; it safeguards and enforces morality; it vouches for the efficiency of ritual and contains practical rules for the guidance of man. . . . Myth is thus a vital ingredient of human civilization."

That science is not everything should not blind us to the fact that it is the very best of what we do have. Just as those who benefit from individual therapy can take pride from the persistent acts of will they exerted along the way, so humankind can take collective pride from the persistent determination to submit to reality therapy that has produced not only the science we now know but also an understanding of how to go about learning more.

Notes on Sources

Chapter 1: Scientific Literacy

The mentioned survey of scientific literacy was described by Jon D. Miller, "Some New Measures of Scientific Illiteracy," Annual Meeting of the American Association for the Advancement of Science, 25–30 May 1986, Philadelphia.

The cited titles are from Christine Russell, "America's Scientific Illiterates," *Washington Post,* 2 June 1986, p. A1; George Gerbner, "Americans' Disdain of Science," *Roanoke Times & World-News,* 4 May 1987; Barbara J. Culliton, "The Dismal State of Scientific Literacy," *Science* 243 (1989): 600; James Krieger, "Past Decade Shows No Gain in U.S. Science Literacy," *Chemical & Engineering News,* 30 January 1989, pp. 24–26. The Sigma Xi lectures are reported in *American Scientist* 78 (1990): 86.

Project 2061 is briefly described by Ward Worthy, "Scientific Literacy: Sweeping Changes in Teaching Urged," *Chemical & Engineering News,* 27 February 1989, p. 4, and "AAAS Offers Guidelines to Combat Scientific Illiteracy Problem," *Chemical & Engineering News,* 13 March 1989, pp. 22–24. The plea for funds came from the Committee for the Scientific Investigation of Claims of the Paranormal (CSICOP) in a flyer dated 16 January 1989.

The rationale underlying the surveys of scientific literacy is discussed by Jon D. Miller, "Scientific Literacy: A Conceptual and Empirical Review," *Daedalus* 112, no. 2 (Spring 1983): 29–48. Details of the questions asked and the method used for scoring are in Miller, "Some New Measures"; and some of these details are also mentioned in Culliton, "Dismal State," and in Krieger, "Past Decade."

Measures

What respondents must say to be regarded as understanding the scientific approach is specified in Miller, "Scientific Literacy," n.22.

That those who understand "the scientific approach" would reject astrology as scientific is stated in Miller, "Scientific Literacy," p. 38. An evenhanded survey of serious astrological investigations is by Geoffrey Dean and Arthur Mather, *Recent Advances in Natal Astrology: A Critical Review, 1900–1976* (Southampton, U.K.: Camelot Press, 1977). Michel Gauquelin, who calls his work astrobiology rather than astrology, describes purely empirical studies that show correlations between human characteristics and planetary positions; see, for example, *Birth-Times: A Scientific Investigation of the Secrets of Astrology* (New York: Hill and Wang, 1983). The results of those studies have been bitterly contested even as the validity of the statistics has been confirmed. See Patrick Curry, "Research on the Mars Effect," *Zetetic Scholar,* no. 9 (1982): 34–53, as well as the discussion on pp. 54–83 in that issue and in *Zetetic Scholar,* no. 10 (1982): 43–81 and *Zetetic Scholar,* no. 11 (1983): 22–33; see also Cornelis De Jager, "Science, Fringe Science and Pseudo-science," *Quarterly Journal of the Royal Astronomical Society* 31 (1990): 31–45; Dennis Rawlins, "sTARBABY," *Fate,* October 1981, pp. 67–98.

The recent history of ideas—and changes in ideas—about how the solar system formed is fully treated by Stephen G. Brush, "Theories of the Origin of the Solar System, 1956–1985," *Reviews of Modern Physics* 62 (1990): 43–112. Critical discussions of details of the possible origin of life on Earth are by A. G. Cairns-Smith, *Seven Clues to the Origin of Life: A Scientific Detective Story* (New York: Cambridge University Press, 1985); John L. Casti, *Paradigms Lost: Images of Man in the Mirror of Science* (New York: William Morrow, 1989); Freeman Dyson, *Origins of Life* (Cambridge: Cambridge University Press, 1985); Robert Shapiro, *Origins: A Skeptic's Guide to the Creation of Life on Earth* (Toronto: Bantam, 1987).

Arguments about the existence of ETI and the possible extraterrestrial provenance of UFOs are discussed by Michael D. Swords, "Science and the Extraterrestrial Hypothesis in Ufology," *Journal of UFO Studies,* n.s., 1 (1989): 67–102.

Achieving Scientific Literacy

The suggestion that about sixteen credit hours of science would give the hoped-for understanding of science was reported by James Krieger, "Science Education: Comprehensive Approach Urged," *Chemical & Engineering News,* 14 May 1990, pp. 4–5.

The separate rates of literacy in the three components defined to make up scientific literacy are given in Krieger, "Past Decade."

The critique by W. M. Laetsch of what people imagine scientific literacy could accomplish is "A Basis for Better Public Understanding of Science," in *Communicating Science to the Public*, ed. David Evered and Maeve O'Connor (Chichester, U.K.: John Wiley, 1987), pp. 1–18 (quoted remarks are on p. 5).

The quotes about purported consequences of scientific illiteracy are respectively by Jon D. Miller, cited in Culliton, "Dismal State"; Philip Merchant, Jr., "Eliminating Racial Barriers in Science," *Chemical & Engineering News*, 12 February 1990, pp. 28–29; George E. Brown, Jr., "Project 2061: A Congressional View," *Science* 245 (1989): 340; William Hively, "How Much Science Does the Public Understand?," *American Scientist* 76 (1988): 439–44.

The inconclusiveness of evidence for the benefits of fluoridation is reviewed by Bette Hileman, "Fluoridation of Water," *Chemical & Engineering News*, 1 August 1988, pp. 26–42. The quote about drug education is from "Antidrug Bill Boosts Treatment, Education Programs," *Chemical & Engineering News*, 21 November 1988, p. 13. Cultural literacy is the hobbyhorse of E. D. Hirsch, Jr., *Cultural Literacy: What Every American Needs to Know* (New York: Vantage Books, 1988). "In Search of Agricultural Literacy" is the editorial in *Virginia Land and Life*, from the College of Agriculture, Virginia Polytechnic Institute and State University (vol. 1, 1989). *Information Literacy: Revolution in the Library* (New York: Macmillan, 1989) is by Patricia Senn Brevik and E. Gordon Gee. Formation of the Tufts University Environmental Literacy Institute was reported in *Chemical & Engineering News*, 7 May 1990, p. 50.

The Underground Grammarian appears eight times a year (P.O. Box 203, Glassboro, NJ 08028). Some of the pieces from it have been collected in Richard Mitchell, *The Leaning Tower of Babel and Other Affronts by the Undergraduate Grammarian* (Boston: Little, Brown, 1984). Mitchell has also published these noteworthy books about what education properly ought to be and what we have perversely made of it: *Less Than Words Can Say* (Boston: Little, Brown, 1979); *The Graves of Academe* (Boston: Little, Brown, 1981); and *The Gift of Fire* (New York: Simon and Schuster, 1987).

The public relations man quoted is Charlie Russell, who designed the successful campaign in which Trout Unlimited organized the defeat of attempts to build the Two Forks Dam in Colorado. See Chas S. Clifton, "Saving the South Platte: Anatomy of a Trout Unlimited Victory," *Trout*, Summer 1990, pp. 23–32.

That a little knowledge of science can be dangerous is exemplifi[e] in the pseudoscience of Immanuel Velikovsky, as discussed at length Henry H. Bauer, *Beyond Velikovsky: The History of a Public Controversy* (Urbana: University of Illinois Press, 1984), especially chap. 7 and pp. 269ff.

Chapter 2: The So-called Scientific Method

For the history of how the contemporary conventional wisdom about the scientific method came about, see the brief discussions by Casti, *Paradigms Lost,* pp. 15–48, and John Ziman, *An Introduction to Science Studies: The Philosophical and Social Aspects of Science and Technology* (Cambridge: Cambridge University Press, 1984), pp. 13–57, who also give references to fuller treatments.

Prevalence of the belief that the essence of science is its method is remarked, for example, by John C. Burnham, *How Superstition Won and Science Lost: Popularizing Science and Health in the United States* (New Brunswick: Rutgers University Press, 1987), p. 28. This is illustrated, for instance, by discussions of the scientific method in textbooks (see my chap. 6). Stephen G. Brush, in "Should the History of Science Be Rated X?," *Science* 183 (1974): 1164–72, points to a number of prevalent misconceptions about scientific activity, including the emphasis on the scientific method construed as hypothetical-deductive. Again, when fields of academic study seek to improve themselves by becoming scientific, they focus on method and define it as empirical, hypothesis constructing, and testing; for example, in archaeology see Colin Renfrew, Michael J. Rowlands, and Barbara Abbott Seagraves, eds., *Theory and Explanation in Archaeology* (New York: Academic Press, 1982); in marketing science see Paul F. Anderson, "Marketing, Scientific Progress, and Scientific Method," *Journal of Marketing* 47 (1983): 18–31.

Are Chemists Not Scientists?

The review of successes of computational chemistry is by H. F. Schaefer, "Methylene: A Paradigm for Computational Quantum Chemistry," *Science* 231 (1986): 1100–1107.

Erwin Chargaff's acerbic comments about Watson and Crick are in *Essays on Nucleic Acids* (Amsterdam: Elsevier, 1963); "A Quick Climb Up Mount Olympus," *Science* 159 (1968): 1448–49; and a review of Robert Olby, *The Path to the Double Helix,* in *Perspectives in Biology and Medicine* 19 (1976): 289–90. Chargaff's data on the composition of nucleic acids are reviewed in "Chemical Specificity of Nucleic Acids and Mechanism of Their Enzymatic Degradation," *Experientia* 6 (1950):

201–40, from which also comes his quoted equivocal statement about the approximate equalities of the amounts of adenine and thymine and of guanine and cytosine.

For the acceptance of ideas about oscillating chemical reactions, see I. R. Epstein, "Patterns in Time and Space—Created by Chemistry," *Chemical & Engineering News*, 30 March 1987, pp. 24–36.

Is Anyone a Scientist?

The quote from Sir Arthur Eddington is given by G. S. Stent in *The Coming of the Golden Age* (New York: Natural History Press, 1969), p. 31.

Norris S. Hetherington provides examples from astronomy of reported observations that turned out to have no objective basis: "Just How Objective Is Science?," *Nature* 306 (1983): 727–30. Bernard Barber has enumerated instances of the initial scientific resistance to new discoveries that were later accepted: "Resistance by Scientists to Scientific Discovery," *Science* 134 (1961): 596–602.

The scientifically conducted membership drive was lauded in the *Contract Bridge Bulletin*, June 1986, p. 29.

The case that disciplinary differences are aptly described as cultural ones is made in Henry H. Bauer, "Barriers against Interdisciplinarity: Implications for Studies of Science, Technology, and Society (STS)," *Science, Technology & Human Values* 15 (1990): 105–19; and "A Dialectical Discussion of the Nature of Disciplines and Disciplinarity," *Social Epistemology* 4 (1990): 215–27.

The different senses that "stable" has for chemists and for physicists were pointed out by Roald Hoffmann, "Marginalia: Unstable," *American Scientist* 75 (1987): 619–21. That experimentalists and theorists use imagery differently is discussed by Anne Roe, *The Making of a Scientist* (New York: Dodd, Mead, 1952), pp. 142, 146–49.

Constance Holden, "The Politics of Paleoanthropology," *Science* 213 (1981): 737–40, and Roger Lewin, *Bones of Contention: Controversies in the Search for Human Origins* (New York: Simon and Schuster, 1987), remark on the high ratio of speculation to fact in paleoanthropology; and Robert Kirshner has noticed it in astronomy (cited in William Hively, "Science Observer: Superdiscoveries," *American Scientist* 76 [1988]: 15). Peter Sturrock, "An Analysis of the Condon Report on the Colorado UFO Project," *Journal of Scientific Exploration* 1 (1987): 75–100, compares the physicists' expectation that crucial tests can be made with the willingness of astronomers to remain tentative for long periods.

The data about Nobel Prizes for experimental as opposed to theoretical discoveries were taken from the listing in P. Wilhelm, *The Nobel Prize* (Stockholm: Teknowledge, 1983).

Anne Roe, *Making of a Scientist,* pp. 42, 45, compares the ages at which eminent biologists and physicists go into administration and the degree to which they feel pressed for time; and also their relative rates of divorce (p. 57). The political affiliations of scientists in the United States are reported by Everett Carll Ladd, Jr., and Seymour Martin Lipset, "Politics of Academic Natural Scientists and Engineers," *Science* 176 (1972): 1091–1100, and *The Divided Academy: Professors and Politics* (New York: McGraw-Hill, 1975), pp. 116, 121, 344; and in Britain, by A. H. Halsey and M. Trow, *The British Academics* (London: Faber, 1971).

That paleontologists tend to dismiss a single catastrophic cause for extinction of the dinosaurs, and the controversy between the opposing parties, has been written about by Steven M. Stanley, *Extinction* (New York: Scientific American, 1987), and Malcolm W. Browne, "Dinosaur Experts Find Meteorite View Political," *Roanoke Times & World-News,* 9 November 1985, p. NRV3.

Scientists Are Human

Gilbert Highet is the "celebrated humanist educator" I quote; see his preface to *The Art of Teaching* (New York: Vintage Books, 1950).

Genesis of the Myth

A very readable account of the grand, nineteenth-century age of science is in David Knight, *The Age of Science: The Scientific World-view in the Nineteenth Century* (Oxford: Basil Blackwell, 1986). Burnham, *Superstition and Science,* describes the relative effectiveness of nineteenth-century popularizing efforts by scientists, in contrast with the twentieth century, in which popular views about science amount to scientistic superstition.

The complex nature and significance of Copernicus's approach is detailed by Thomas Kuhn, *The Copernican Revolution* (Cambridge: Harvard University Press, 1957). More concise discussions can be read in, for example, F. R. Jevons, *Puzzles and Revolutions: Case Study of the Copernican Revolution* (Waurn Ponds, Australia: Deakin University, 1979); Jerzy Neyman, ed., "Introduction," *The Heritage of Copernicus: Theories "Pleasing to the Mind"* (Cambridge: MIT Press, 1974), pp. 1–22.

For the case that Galileo was persecuted for other reasons than maintaining that the Sun goes around the Earth, see, for instance, Joseph Pitt, *Galileo and the Book of Nature* (Dordrecht: Kluwer, 1991); Pietro Redondi, *Galileo: Heretic* (Princeton: Princeton University Press, 1987); and reviews of the latter by Ron Naylor, "Questions of Heaven and Host," *Nature* 330 (1987): 617–18; Joseph Pitt, "Friends in Holy

Places," *The Sciences,* January/February 1988, pp. 48–56; Richard S. Westfall, "The Case of Galileo," *Science* 237 (1987): 1059.

The Epitome of Science

The Nobel ceremony at which Glashow's award for physics came before the others is described by Sheldon Glashow, who also admits that physicists are, in the literal sense, simpleminded: (with Ben Bova) *Interactions: A Journey through the Mind of a Particle Physicist and the Matter of This World* (New York: Warner, 1988), pp. xi, 3, 279–80.

Wilhelm, *Nobel Prize,* published for the 150th anniversary of Nobel's birth, also lists winners in the sequence physics, chemistry, physiology or medicine, literature, peace (and economic science, since 1969). Prominent physicists who have been fooled by mediums and psychics are mentioned in Paul Kurtz, ed., *A Skeptic's Handbook of Parapsychology* (Buffalo, N.Y.: Prometheus, 1985), especially chaps. 6, 10, and 14.

For the role of chemists in the discovery of nuclear fission, see, for instance, Stephen G. Brush, *The History of Modern Science: A Guide to the Scientific Revolution, 1800–1950* (Ames: Iowa State University Press, 1988), pp. 351–52, 362.

The qualifications of physicists to improve economic forecasting were commented on by a biologist and a psychologist in letters to *Science* 246 (1989): 10.

For estimated costs of the Superconducting Super-Collider, see Kim A. McDonald, "House Backs Spending for Superconducting Super Collider, but Sets $5-billion Limit on Federal Share," *Chronicle of Higher Education,* 9 May 1990, p. A24.

From Myth to Ideal

The Nobelist who gave assurance of his own integrity is David Baltimore; the quote is in Wil Lepkowski, "Academic Values Tested by MIT's New Center," *Chemical & Engineering News,* 15 March 1982, pp. 7–12.

As this book is prepared for publication, final judgments on the Imanishi-Kari affair are still pending. The congressional hearings in May 1989 were chaired by Congressman John Dingell of the Subcommittee on Oversight of the House Committee on Energy and Commerce. That misconduct had been proven was stated in a draft report by the Office of Scientific Integrity of the National Institutes of Health. See Pamela S. Zurer, "Scientific Whistleblower Vindicated," *Chemical & Engineering News,* 8 April 1991, pp. 35–40.

Other recent scandals include: a psychologist who falsified data about the effects of stimulant drugs on hyperactive children (see Gregory

Byrne, "Breuning Pleads Guilty," *Science* 242 [1988]: 27–28); ophthal-mologists who violated regulations governing research with human sub-jects and withheld information that a treatment in which they had financial interest was ineffective (see William Booth, "Hospital Faulted for Dry Eye Study," *Science* 243 [1989]: 1000); a geologist who lied about the provenance of fossils (see Roger Lewin, "The Case of the 'Misplaced' Fossils," *Science* 244 [1989]: 277–79); psychiatrists who fal-sified clinical data (see Marcia Barinaga, "NIMH Assigns Blame for Tainted Studies," *Science* 245 [1989]: 812); a biology professor at Pur-due University who stole information from a manuscript he was re-viewing (see Pamela S. Zurer, "NIH Panel Strips Researcher of Funding after Plagiarism Review," *Chemical & Engineering News*, 7 August 1989, pp. 24–25). And there is still controversy over whether an eminent British psychologist made up some or all of his data on identical twins (see Constance Holden, "Rehabilitation for Burt?," *Science* 251 [1991]: 27); and about whether the first isolation of the AIDS virus came from stolen samples (see D.P.H., "More Woes for Gallo," *Science* 251 [1991]: 152).

Chapter 3: How Science Really Works

Cooperative Action in Science: The Jigsaw Puzzle

Michael Polanyi describes the jigsaw puzzle analogy to scientific ac-tivity in "The Republic of Science: Its Political and Economic Theory," *Minerva* 1 (1962): 54–73.

John Ziman has developed in great detail the picture of scientific activity as a striving to attain rational consensus; see, for example, *Public Knowledge: An Essay concerning the Social Dimensions of Science* (Cam-bridge: Cambridge University Press, 1968), p. 9. He remarks that the primary literature is information rather than knowledge (p. 120); and also that frontier physics may be 90 percent wrong (*Reliable Knowledge: An Exploration of the Grounds for Belief in Science* [Cambridge: Cambridge University Press, 1978], p. 40).

That most scientific papers are not cited at all, or only a handful of times, whereas a few are cited very heavily, is one aspect of the unequal distribution of talent or achievement in science that Derek de Solla Price discusses in one of his seminal articles, "Galton Revisited," *Little Science, Big Science . . . and Beyond* (New York: Columbia University Press, 1986), chap. 2. Specific studies showing lack of citation of much of the scientific literature are in Henry W. Menard, *Science: Growth and Change* (Cambridge: Harvard University Press, 1971), p. 99, and J. R. Cole and

S. Cole, *Social Stratification in Science* (Chicago: University of Chicago Press, 1973), p. 228.

The Scientific Method versus the Filter

Derek de Solla Price, *Science since Babylon* (New Haven: Yale University Press, 1975), describes and discusses the exponential growth of science from an origin in the seventeenth century. A good account has also been given by John Ziman, *The Force of Knowledge: The Scientific Dimension of Society* (Cambridge: Cambridge University Press, 1976), chap. 3.

Useful, albeit early, reviews of the controversy over claims of cold fusion are by Sharon Begley, "The Race for Fusion," *Newsweek,* 8 May 1989, pp. 48–54; Ron Dagani, "Hopes for Fusion Diminish as Ranks of Disbelievers Swell," *Chemical & Engineering News,* 22 May 1989, pp. 8–20; Michael D. Lemonick, "Fusion or Illusion?," *Time,* 8 May 1989, pp. 72–77. The early book by F. David Peat, *Cold Fusion: The Making of a Scientific Controversy* (Chicago: Contemporary Books, 1989), adds nothing of note to those reviews. More comprehensive but hastily assembled is *Too Hot to Handle: The Race for Cold Fusion* (Princeton: Princeton University Press, 1991), by Frank Close. Well worth reading is Eugene F. Mallove, *Fire from Ice: Searching for the Truth behind the Cold Fusion Furor* (New York: John Wiley & Sons, 1991).

Science and technology studies is also characterized by diversity. To varying degrees we bring to STS biases learned within our main discipline. There is no complete agreement within STS, for example, over the degree to which scientific knowledge approaches objectivity by contrast to reflecting the social interests of scientists. Indeed, there remain a few—perhaps particularly philosophers—who believe that a fairly straightforward unity of method characterizes science. This book is written from the viewpoint that no simple unity characterizes the practice of science at all times but that much scientific knowledge approaches quite close to reflecting salient characteristics of the objective, real world.

Thomas Kuhn's classic, *The Structure of Scientific Revolutions* (Chicago: University of Chicago Press, 1962), is usually referred to in the revised version published in 1970. A relatively recent critique of Kuhn's "irrationalism" is by David Stove, *Popper and After: Four Modern Irrationalists* (Oxford: Pergamon, 1982).

The classic modern dismissal of a variety of pseudosciences is Martin Gardner, *Fads and Fallacies in the Name of Science* (New York: Dover, 1957). Persistent attacks on such things are featured in the quarterly *Skeptical Inquirer.*

The classic piece on resistance by scientists to new discovery is Barber, "Resistance by Scientists," concise but never superseded.

The paper claiming reactions at extreme dilution was by E. Davenas et al., "Human Basophil Degranulation Triggered by Very Dilute Antiserum against IgE," *Nature* 333 (1988): 816–18. Salient aspects of the ensuing controversy are argued by Jacques Benveniste in *Nature* 334 (1988): 291 and *Science* 241 (1988): 1028; in John Maddox, James Randi, and Walter W. Stewart, " 'High-Dilution' Experiments a Delusion," *Nature* 334 (1988): 287–90; and in Robert Pool, "Unbelievable Results Spark a Controversy," *Science* 241 (1988): 407 and "More Squabbling over Unbelievable Results," ibid.: 658.

For details of Gauquelin's studies and the controversy, see the references on p. 154. The studies carried out in the Princeton Engineering Anomalies Research program are summarized in Robert G. Jahn, "The Persistent Paradox of Psychic Phenomena: An Engineering Perspective," *Proceedings of the Institute of Electrical and Electronics Engineers* 70 (1982): 136–70; and Robert G. Jahn and Brenda J. Dunne, *Margins of Reality: The Role of Consciousness in the Physical World* (San Diego: Harcourt Brace Jovanovich, 1987).

The technically sophisticated efforts of the Academy of Applied Science at Loch Ness are described in Robert H. Rines, Charles W. Wyckoff, Harold E. Edgerton, and Martin Klein, "Search for the Loch Ness Monster," *Technology Review,* March-April 1976, pp. 25–40. What differentiates the Loch Ness investigations from scientific investigations is discussed in Henry H. Bauer, *The Enigma of Loch Ness: Making Sense of a Mystery* (Urbana: University of Illinois Press, 1986).

Martin Gardner wrote "The Hermit Scientist," *Antioch Review* 10 (1950): 447–57. A full analysis of Velikovsky's pseudoscience is in Bauer, *Beyond Velikovsky.* The polywater episode is reviewed in Felix Franks, *Polywater* (Cambridge: MIT Press, 1981); N-rays, in Mary Jo Nye, "N-rays: An Episode in the History and Psychology of Science," *Historical Studies in the Physical Sciences* 11, pt. 1 (1980): 125–56.

Regarding Albert Szent-Gyorgyi, I was informed chiefly by the biography by Ralph W. Moss, *Free Radical* (New York: Paragon, 1988). Quite a number of books have been written about Wilhelm Reich, almost all of them by people predisposed in his favor. I think the most illuminating and evenhanded is by Colin Wilson, *The Quest for Wilhelm Reich* (Garden City, N.Y.: Anchor/Doubleday, 1981).

Chapter 4: Other Fables about Science

How wrongheaded beliefs about science complicate public controversies is illustrated in *Beyond Velikovsky* and *The Enigma of Loch Ness*, my book-length studies of arguments on the fringes of science.

Science Deals in Facts

The lack of evidence underlying the earlier belief in parity conservation is mentioned by Jeremy Bernstein, *A Comprehensible World* (New York: Random House, 1967), p. 52.

That much of science is 99.9 percent certain was said by Philip Abelson under provocation from Velikovsky's supporters in a letter to W. Sizemore, quoted in Alfred de Grazia, ed., *The Velikovsky Affair* (New York: University Books, 1966), p. 188. The rhetoric by which the authority of science is invoked while lip service is paid to the fallibility of science has been analyzed in Bauer, *Beyond Velikovsky*, chaps. 8 and 14. For example, Carl Sagan called improbable hypotheses "untenable," in *Broca's Brain* (New York: Random House, 1979), p. 99.

An excellent discussion of the theory-ladenness of fact is in Harold I. Brown, *Perception, Theory and Commitment: The New Philosophy of Science* (Chicago: University of Chicago Press, 1979).

Among those who have stated baldly that evolution is a fact, not a theory, are L. Sprague and Catherine C. De Camp, *Spirits, Stars and Spells* (New York: Canaveral, 1966), p. 292; S. J. Gould, "Evolution as Fact and Theory," *Discover*, May 1981, pp. 34–37; A. Kornberg, quoted in "Science People: The Elite Meet," *Discover*, April 1981, p. 62; Richard E. Leakey, quoted in Cheryl M. Fields, "Anthropologists Continue Debate on Man's Origins but Agree on Dangers Posed by Creationism," *Chronicle of Higher Education*, 14 April 1982, p. 5; Carl Sagan, *Cosmos* (New York: Random House, 1980), p. 27. Evolutionists also find it difficult not to read "progress," "improvement," "increased complexity," or something of that sort into the fossil record, though the *"facts,"* of course, entail no such judgment.

The creationist criticism of science for changing its estimate of the Earth's age is cited in Shapiro, *Origins*, p. 260. Much criticism of the creationists seems to stem from the conviction that creationist belief is harmful: to those who hold it, or to their children, or to everyone's children, if creationism gets mentioned equally with evolution in the schools; or, in the most sweeping sense, because it is dangerous for society when irrational beliefs flourish. That viewpoint is characteristic, for instance, of the Committee for the Scientific Investigation of Claims of the Paranormal (or of the American Humanist Association, out of which CSICOP was founded). Thus, CSICOP investigators have perceived possible danger from psychic belief three times as often as possible benefit, while other people either have seen benefits outweighing dangers or have estimated the danger as no more than twice as great as the benefit; see Roger Klare, "Ghosts Make News," *Skeptical Inquirer* 14 (Summer 1990): 363–71.

That the published scientific literature is not equally open to everyone is pointed out by Ziman, *Introduction to Science Studies,* pp. 68, 178. Examples of Velikovsky's misleading by quoting out of context are in Bauer, *Beyond Velikovsky.*

Recommendations concerning cholesterol levels have been shown to stem from particular contexts that may not apply to the population as a whole; see Thomas J. Moore, "The Cholesterol Myth," *Atlantic,* September 1989, pp. 37–70.

Successful Prediction Proves a Theory Right

Stephen G. Brush has discussed the role that prediction plays (or does not play) in the evaluation of scientific theories in "Prediction and Theory Evaluation: The Case of Light Bending," *Science* 246 (1989): 1124–29, and "Prediction and Theory Evaluation: Alfven on Space Plasma Phenomena," *Eos* 71, no. 2 (1990): 19–33. Bauer, *Beyond Velikovsky,* pp. 86ff., has shown why Velikovsky's predictions were no support for his cosmic scenario.

Science Is (or Should Be) Open-minded

David Hull has written at length about the analogy between biological evolution and the progress of science in *Science as a Process: An Evolutionary Account of the Social and Conceptual Development of Science* (Chicago: University of Chicago Press, 1988).

The distinction among known, known unknown, and unknown unknown is from Peter J. Denning, "The Science of Computing: Blindness in Designing Intelligent Systems," *American Scientist* 76 (1988): 118–20.

The quote from Fred Hoyle is in *The Nature of the Universe* (New York: Harper and Row, 1960), p. 134; that from Carl Sagan, in *Broca's Brain,* p. xv.

Factors influencing the resistance by scientists to new discoveries are pointed out by Barber, "Resistance by Scientists."

Chesterton's bon mot about open-mindedness is in his *Autobiography* (New York: Sheed and Ward, 1936), p. 229. The *reductio ad absurdum* of claims that science should be open-minded is discussed in Bauer, *Beyond Velikovsky,* pp. 281–82.

For how scientists choose problems to work on, see, for example, Ziman, *Public Knowledge,* pp. 98ff.; and Peter B. Medawar, *Advice to a Young Scientist* (New York: Harper and Row, 1979). Misunderstandings occasioned by Medawar's description of science as *The Art of the Soluble* (London: Methuen, 1967), interpreted by some as the choice of easy problems, are spoken to in *Pluto's Republic* (Oxford: Oxford University Press, 1982), pp. 2–3. Often the choice is not made primarily by scientists; in industry, for example, scientists display what has been aptly described by Gordon Tullock as "induced curiosity" (*The Organization of Inquiry* [Durham: Duke University Press, 1966]). That there is always risk in making the choice, and that scientists of different temperament choose differently, is implicit in Szent-Gyorgyi's aphorism "It's much more exciting not to catch a big fish than not to catch a little one" (quoted in Gregory Byrne, "Random Samples," *Science* 241 [1988]: 1165).

Scientists Should . . .

Robert Millikan's notes are analyzed in Gerald Holton, "Electrons or Subelectrons? Millikan, Ehrenhaft and the Role of Preconceptions," *History of Twentieth-Century Physics*, ed. C. Weiner (New York: Academic Press, 1977), pp. 266–89; and by Allan Franklin, "Millikan's Published and Unpublished Data on Oil Drops," *Historical Studies in the Physical Sciences* 11, no. 2 (1981): 185–201. The charge of misconduct is leveled in William Broad and Nicholas Wade, *Betrayers of the Truth: Fraud and Deceit in the Halls of Science* (New York: Simon and Schuster, 1982), pp. 33–35.

The letter urging that Velikovsky's predictions be credited was from V. Bargmann and Lloyd Motz, "On the Recent Discoveries concerning Jupiter and Venus," *Science* 138 (1962): 1350. Why credit should *not* have been given *by scientists* is argued in Bauer, *Beyond Velikovsky*, pp. 41ff., 285ff.

The lack of citation of Hannes Alfvén's work is treated in Brush, "Origin of the Solar System."

Broad and Wade, *Betrayers of the Truth*, write in sweeping fashion about improper conduct by scientists. For critiques of their views, see the reviews by Henry H. Bauer, "Betrayers of the Truth: A Fraudulent and Deceitful Title from the Journalists of Science," *4S Review* 1, no. 3 (1983): 17–23; Walter Gratzer, "Trouble at t'Lab," *Nature* 302 (1983): 774–75; David Joravsky, "Unholy Science," *New York Review of Books*, 13 October 1983, pp. 3–5; John Ziman, "Fudging the Facts," *Times Literary Supplement*, 9 September 1983, p. 955; and Norton D. Zinder, "Fraud in Science: A Scientist's View," *Science '83*, January-February 1983, pp. 94–95.

Science Is Self-correcting

The quotation on the Church at the time of the Reformation is from De Lamar Jensen, *Reformation Europe: Age of Reform and Revolution* (Lexington, Mass.: Heath, 1981), p. 14.

James D. Watson's autobiographical account is *The Double Helix: A Personal Account of the Discovery of the Structure of DNA* (New York: Atheneum, 1968). An interesting edition that includes other historical accounts, reviews of the book, and evaluative comments is by Gunther S. Stent (New York: W. W. Norton, 1980). Horace Freeland Judson's comment is in *The Eighth Day of Creation: Makers of the Revolution in Biology* (New York: Simon and Schuster, 1979), p. 69; Max F. Perutz's is in "How the Secret of Life Was Discovered," *Daily Telegraph* (London), 27 April 1987, reprinted as "Discoverers of the Double Helix," in *Is Science Necessary? Essays on Science and Scientists* (New York: E. P. Dutton, 1989), pp. 181–83.

Other works that could be considered part of the modern genre of scientific docunovels include: Gary Taubes, *Nobel Dreams: Power, Deceit, and the Ultimate Experiment* (New York: Random House, 1986); Glashow, *Interactions;* Natalie Angier, *Natural Obsessions: The Search for the Oncogene* (Boston: Houghton-Mifflin, 1988); David H. Clark, *The Quest for SS433* (New York: Viking, 1985); Jeff Goldberg, *Anatomy of a Scientific Discovery* (New York: Bantam, 1988); Stephen S. Hall, *Invisible Frontiers: The Race to Synthesize a Human Gene* (New York: Atlantic Monthly, 1987); Robert M. Hazen, *The Breakthrough: The Race for the Superconductor* (New York: Summit, 1988); Robert Kanigel, *Apprentice to Genius: The Making of a Scientific Dynasty* (New York: Macmillan, 1986); Tracy Kidder, *The Soul of a New Machine* (Boston: Little, Brown, 1981); Charles E. Levinthal, *Messengers of Paradise: Opiates and the Brain* (New York: Anchor/Doubleday, 1988); Roger Lewin, *Bones of Contention;* David M. Raup, *The Nemesis Affair: A Story of the Death of Dinosaurs and the Ways of Science* (New York: W. W. Norton, 1986); Ed Regis, *Who Got Einstein's Office: Eccentricity and Genius at the Institute for Advanced Study* (Reading, Mass.: Addison-Wesley, 1987); Bruce Schechter, *The Path of No Resistance: The Story of the Revolution in Superconductivity* (New York: Simon and Schuster, 1988); Solomon H. Snyder, *Brainstorming: The Science and Politics of Opiate Research* (Cambridge: Harvard University Press, 1989); Robert Teitelman, *Gene Dreams: Wall Street, Academia, and the Rise of Biotechnology* (New York: Basic Books, 1989); Nicholas Wade, *The Nobel Duel: Two Scientists' 21-Year Race to Win the World's Most Coveted Research Prize* (Garden City, N.Y.: Anchor/Doubleday, 1981).

Great Scientists Can Speak for Science

The unfounded statement about double bonds was made by George Wald, 1967 joint Nobelist for physiology or medicine, speaking (more than a decade later) at Virginia Polytechnic Institute and State University. William Shockley, 1956 joint Nobelist in physics, embarrassed many scientists by his simpleminded crusade to deter unintelligent races from reproducing.

That the same discovery appears often to be made independently by several people at about the same time has been the subject of classic papers by Robert K. Merton, "Priorities in Scientific Discovery," *American Sociological Review* 22, no. 6 (1957): 635–59, and "Singletons and Multiples in Scientific Discovery," *Proceedings of the American Philosophical Society* 105, no. 5 (1961): 470–86; and by Price, *Science since Babylon*, pp. 59ff.

Erwin Chargaff's aphorism is in "A Quick Climb Up Mount Olympus."

Chapter 5: Imperfections of the Filter

Objectivity versus Consensus

For exposition and discussion of "constructivist" and "relativist" approaches, see Karin D. Knorr-Cetina and Michael Mulkay, *Science Observed* (London: Sage, 1983); and Thomas F. Gieryn, "Relativist/ Constructivist Programmes in the Sociology of Science," *Social Studies of Science* 12, no. 2 (May 1982): 279–98, with responses and replies on pp. 299–336. Paul Feyerabend is perhaps the most notorious propagandist of the view that scientific knowledge is nothing special; see, for instance, *Against Method* (London: New Left Books, 1975). The self-contradictions into which people fall when they try to avoid a realist stance in talking about science has been incisively (and delightfully) shown by Olga Amsterdamska, "Surely You Are Joking, Monsieur Latour!," *Science, Technology & Human Values* 15, no. 4 (Autumn 1990): 495–504.

John Ziman's statement about rational consensus is in *Public Knowledge*, p. 9. Richard M. Burian talked of "reality therapy" in "Discipline Transformation: Social Construction, Integration of Nature, and Reality Therapy," a paper delivered at the closing plenary session of the Summer Conference on the History, Philosophy, and Social Studies of Biology, at Blacksburg, Virginia, in 1987.

Conflicts of Interest

The quotations and examples in this section are from: Daniel S. Greenberg, "The Baltimore Case," *Issues in Science and Technology,* Winter 1989–90, pp. 25–26; Pamela Zurer, "Proposed Ethical Guidelines Viewed as Overkill," *Chemical & Engineering News,* 27 November 1989, pp. 42, 44; David Baltimore, "Baltimore's Travels," *Issues in Science and Technology,* Summer 1989, pp. 48–54; E.M. [Eliot Marshall], "The Florida Case: Appearances Matter," *Science* 248 (1990): 153; Marcia Barinaga, "Biotechnology on the Auction Block," *Science* 247 (1990): 906–8; Eliot Marshall, "When Commerce and Academe Collide," *Science* 248 (1990): 152–56; David L. Wheeler, "Health Secretary Kills Guidelines to Bar Abuse among NIH Grantees," *Chronicle of Higher Education,* 10 January 1990, pp. A1, A22; Joseph Palca, "NIH Conflict-of-Interest Guidelines Shot Down," *Science* 247 (1990): 154; Deborah Runkle, "Conflict of Interest in Science," *Science* 246 (1989): 1177; J.P., "Some of the Voices from the Chorus of Protest," *Science* 247 (1990): 155.

The study of Health Stop services was reported on CNN's "Headline News" on 11 April 1990 and in Ron Winslow, "Physicians Offered Incentives at Clinics Prescribed Far More Lab Tests, X-rays," *Wall Street Journal,* 12 April 1990, p. B4.

Support of Science

Derek de Solla Price discusses the very unequal distribution of scientific achievements in chapter 2 of the revised edition of *Little Science, Big Science.*

The AAU's unwillingness to demand virtue of its members was reported in Wil Lepkowski, "Debate Persists on University Lobbying," *Chemical & Engineering News,* 13 April 1987, p. 28.

Scientific Communities

The N-ray episode is discussed in Nye, "N-rays."

The industrial research that misled itself by not making data openly available is mentioned in Marcia Barinaga, "The Missing Crystallographic Data," *Science* 245 (1989): 1179–81.

The Strategic Defense Initiative (SDI, or Star Wars) was in fact given an unsolicited review by a committee of the American Physical Society. The lengthy full report is in *Reviews of Modern Physics* 59, no. 3, pt. 2 (July 1987): S1–201; its tenor is accurately reflected by Colin Norman's summary "Doubt Cast on Laser Weapons," *Science* 236 (1987): 509–10: "An American Physical Society report says major technical advances and at least another decade of research will be required to determine whether directed energy weapons will work."

Medawar's remark is quoted by Perutz, *Is Science Necessary?,* p. 196.

The lack of theoretical or empirical basis for long-used equations for speeds of electrochemical reactions is discussed in Henry H. Bauer, "The Electrochemical Transfer-Coefficient," *Journal of Electroanalytical Chemistry* 16 (1968): 419–32.

Bias and Progress in Science

The reaction to studies correlating cancer with fluoride is described by Bette Hileman, "Fluoridation of Water," *Chemical & Engineering News*, 7 May 1990, p. 4.

That science benefits when scientists are humanly heterogeneous, simply because science is a consensual activity that aims toward objectivity, may seem obvious; still, the most closely reasoned, explicit argument to that effect that I have seen was not in the scholarly literature but in a term paper by Michelle A. Horvath, "One Science, Once and for All: A Discussion of Feminist Critiques of Science and the Problems Facing Modern Science" (Spring 1989, Virginia Polytechnic Institute and State University, HUM 4304—Contemporary Issues in Humanities, Science, and Technology).

Chapter 6: Consequences of Misconception

Frontier Science and Textbook Science

The fastest-spinning object that turned out to be a TV signal was reported in Kim A. McDonald, "Big Astronomical 'Discovery' Turns Out to Be Interference Signal from Television," *Chronicle of Higher Education*, 28 February 1990, pp. A1, A11.

Charges against Mendel, Newton, Ptolemy, and other famous scientists are made in Broad and Wade, *Betrayers of the Truth*. The deliberate fraud by Breuning is described in Byrne, "Breuning Pleads Guilty."

Alvin M. Weinberg was among the first to emphasize the impossibility of demonstrating *zero* risk from such factors as radiation: "Science and Trans-science," *Minerva* 10 (1972): 209–22.

The burgeoning of alternative medicine in Britain is illustrated by the writings of Jan de Vries in the series *By Appointment Only* (Edinburgh: Mainstream, 1988), with such individual titles as *Cancer and Leukemia: An Alternative Approach, Do Miracles Exist?,* and *Traditional Home and Herbal Remedies*. Those ventures have been described with respect in at least one national newspaper: Jean Smith, "The Natural Ways of a Dutch Healer," *Scotsman*, 19 April 1985, p. 8.

Fred Hoyle's statement is from *The Quasar Controversy Resolved* (Cardiff: University College Cardiff Press, 1981), p. 7.

For criticism that David Hull fails to do justice to the intellectual aspect of science, see reviews of his book *Science as a Process* by, for example, J. Maynard Smith, *Science* 242 (1988): 1182–83.

The need to distinguish frontier from textbook science is asserted in Henry H. Bauer, "Frontier Science and Textbook Science," *Science &*

Technology Studies 4, no. 3/4 (1986): 33–34. Disciplinary barriers to STS are discussed in Bauer, "Barriers against Interdisciplinarity."

Robert Pool, "Superconductivity: Is the Party Over?," *Science* 244 (1989): 914–16, asks the question; C. W. Chu and D. U. Gubser et al., *Science* 245 (1989): 111–12, answer in the negative. But as far as the media and the public are concerned, the party *is* over.

Science and Technology

For how scientists choose tractable but worthwhile problems, see, for instance, Medawar, *Advice to a Young Scientist,* and Ziman, *Public Knowledge.*

Some scientists who feel the urge to jump into the unknown unknown—joining in the International Society for Cryptozoology, for example—are generally careful to distinguish such activity from their professional scientific work, for the good reason that failure to do so can be professionally damaging (see Bauer, *Enigma of Loch Ness,* pp. 121ff). The Society for Scientific Exploration and its *Journal of Scientific Exploration* were founded by scientists explicitly wishing to provide a critically sophisticated forum for discussion of matters outside the mainstream and too tentative for the usual disciplinary conferences or journals. Another periodical with similar intent is *Speculations in Science and Technology.* A collection of such items has been published by I. J. Good, *The Scientist Speculates: An Anthology of Partly-Baked Ideas* (New York: Basic Books, 1963).

How peer review discriminates against the brilliant has been written about by, for instance, Thomas Gold, "New Ideas in Science," *Journal of Scientific Exploration* 3 (1989): 103–12; Bill Lawren, "Harmon Craig: Stalking Excellence, Leaving Controversy in His Wake," *The Scientist,* 17 April 1989, pp. 1, 18, 19; and Albert Szent-Gyorgyi, "Dionysians and Apollonians," *Science* 176 (1972): 966.

Price, *Little Science, Big Science,* chap. 2, emphasizes the skewed distribution of scientific achievement and its implications.

Diverting funds from the purported project to something more interesting but risky is a common device, but publicly admitted only rarely; see, for instance, Richard A. Muller, "Innovation and Scientific Funding," *Science* 209 (1980): 880–83. Another scheme is to request funds for work already performed, using the money for what one actually wants to do and incorporating whatever results eventuate into the proposal for the following period. But that runs the risk that one might get scooped by keeping results unpublished for a year or more, and so this scheme has been used in oppressively bureaucratic societies more than elsewhere. See, for example, Eric Ashby, *Scientist in Russia* (Harmondsworth, U.K.: Penguin, 1947).

That Stanley Pons and Martin Fleischmann asked for no research support in the early stages of their quest for cold fusion is reported by Begley, "Race for Fusion"; Dagani, "Hopes for Fusion"; and Lemonick, "Fusion or Illusion?"

The quotes relating to the National Science Foundation are from "User Survey Okays NSF Peer Review System," *Chemical & Engineering News*, 30 May 1977, pp. 16, 21; and Eliot Marshall, "A Fast Track for High-Risk Science," *Science* 244 (1989): 764.

Erwin Chargaff compares goal-oriented research to alchemy in *Voices in the Labyrinth* (New York: Seabury, 1977), pp. 89, 125.

The quote about dictatorships and poor science is from Daniel E. Koshland, Jr., "Communism, Capitalism, and Dissent," *Science* 245 (1989): 109.

Philipp Lenard, 1905 Nobelist, wrote a textbook of nonrelativistic Nazi physics, *Deutsche Physik*, 2d ed., 4 vols. (Munich: J. F. Lehmann, 1938).

On the matter of questions that in reality have no scientific answer, see Weinberg, "Science and Trans-science."

Instances of pure science sparked by technological developments are in John P. McKelvey, "Science and Technology: The Driven and the Driver," *Technology Review,* January 1985, pp. 38–47. Many aspects of the complicated relationship between science and technology are discussed by Price, *Science since Babylon,* including the ancestry of astronomical models and time-keeping devices and the fact that technology, insofar as it is based on science at all, is based on *old* rather than on *new* science.

Chargaff's remark that science makes the men is in "A Quick Climb Up Mount Olympus."

Social and Behavioral Science

The change in name of the British Social Research Council was reported in *Chronicle of Higher Education,* 14 September 1983, p. 27.

For a recent account of the lack of consensus about how to help disadvantaged children, see Constance Holden, "Head Start Enters Adulthood," *Science* 247 (1990): 1400–1402. For a recent argument over the efficacy of Prohibition, see John C. Burnham, "Drug Decriminalization," *Science* 246 (1989): 1102. For one polemic among psychotherapies, see Albert Ellis, *Why Some Therapies Don't Work* (Buffalo, N.Y.: Prometheus, 1989).

For a recent discussion of the weakness of reality therapy in social science, see Michael T. Ghiselin, *Intellectual Compromise: The Bottom Line* (New York: Paragon House, 1989), especially pp. 65ff. For the misguided denigration of commonsense knowledge by sociologists, see James A. Mathisen, "A Further Look at 'Common Sense' in Introductory Sociology," *Teaching Sociology* 17 (1989): 307–15.

The nature of professional evaluation in different disciplines is treated in Josef Martin (a pen name of Henry H. Bauer), *To Rise above Principle: The Memoirs of an Unreconstructed Dean* (Urbana: University of Illinois Press, 1988), pp. 116–19. The relative consistency of refereeing of journal articles in physics and psychology is mentioned in Broad and Wade, *Betrayers of the Truth*, pp. 102–3.

Burnham, *Superstition and Science*, pp. 112, 299, notes that introductory psychology texts are invariably preoccupied with the scientific method, whereas texts in the natural sciences are not. I quote from the following introductory sociology texts: Glenn M. Vernon, *Human Interaction: An Introduction to Sociology* (New York: Ronald Press, 1965), pp. 20, 35, 36; John Perry and Erna Perry, *The Social Web: An Introduction to Sociology*, 2d ed. (San Francisco: Canfield Press, 1976), p. 29; Ian Robertson, *Sociology* (New York: Worth, 1977), p. 8.

For the attempt to guide archaeology scientifically, see Colin Renfrew, Michael J. Rowlands, and Barbara Abbott Seagraves, eds., *Theory and Explanation in Archaeology* (New York: Academic Press, 1982).

The quote about the unscientific nature of the social sciences is from Ernest Gellner, "The Scientific Status of the Social Sciences," *International Social Science Journal* 36 (1984): 567–86. See also Stanislav Andreski, *Social Sciences as Sorcery* (London: Andre Deutsch, 1972); Stephen Park Turner and Jonathan H. Turner, *The Impossible Science* (Newbury Park, Calif.: Sage, 1990).

The meeting of the Society for Social Studies of Science at which *scientists* were asked to raise their hands took place on 2–4 November 1979 in Washington, D.C.

Chapter 7: In Praise of Science

The quote about "Science-Ads" is from Thomas F. Gieryn and Elizabeth Hunt, that about "succession rites" from P. G. Abir-Am, both in "Abstracts of the 1987 Meeting of the Society for Social Studies of Science," *Science & Technology Studies* 5 (1987): 74–76.

Vive la Différence!

Ziman, in *Public Knowledge* and *Reliable Knowledge*, explicates manifold details of the active striving for consensus in science.

Science Is Not Scientism

C. P. Snow's *Two Cultures: And a Second Look* (New York: Mentor, 1964 and several other editions), on the cultural aspects of science, remains a classic. Among the few who have more recently emphasized the cultural in contrast to the material benefits of science is Leon M. Lederman, "The Value of Fundamental Science," *Scientific American* 251 (November 1984): 40–47.

Scientific Literacy and Education in Science

The quotation about those who have no understanding of the world is from Roald Hoffmann's Priestley Medal Address, which appears in *Chemical & Engineering News,* 23 April 1990, pp. 25–29.

The Scientific Method as an Ideal

For the patterned, nonanalytic way of thought that comes naturally to humans, see R. E. Nisbett and L. Ross, *Human Inference: Strategies and Shortcomings of Social Judgment* (Englewood Cliffs, N.J.: Prentice-Hall, 1980); and Howard Margolis, *Patterns, Thinking, and Cognition* (Chicago: University of Chicago Press, 1987). Some of the examples given in the text are drawn from Daryl Bem, "The Intuitive Science of Everyday Information Processing: Just How Bad Are We?," Eighth Annual Meeting of the Society for Scientific Exploration, Boulder, Colo-

The birthday problem is worked in, for example, John G. Kemeny, J. Laurie Snell, and Gerald L. Thompson, *Introduction to Finite Mathematics,* 3d ed. (Englewood Cliffs, N.J.: Prentice-Hall, 1974), pp. 93–94.

The quotation defining myth is attributed to Bronislaw Malinowski by Rollo May, "The Therapist and the Journal into Hell," *National Forum* 69, no. 2 (Spring 1989): 33–37.

Further Reading

I hope to have shown that the puzzle and filter model gives a usable understanding of what goes on in science and how that meshes with other societal activities or institutions. I hope too to have demonstrated that popular misconceptions are rife, notably about the scientific method. More detail about these matters can be gleaned from the references cited in the Notes on Sources; here I mention only a small number of exceptionally noteworthy writings.

Thomas Kuhn's *Structure of Scientific Revolutions,* 2d (enlarged) ed. (Chicago: University of Chicago Press, 1970) is a classic, a milestone in scholarly recognition that science is not just a matter of abstract intellection. An earlier, popularly written exposition by Anthony Standen, *Science Is a Sacred Cow* (New York: E. P. Dutton, 1950), discusses common illusions about science, but several parts of that book are now clearly dated. A classic, and not at all dated, on how popular opinion distorts realities about science is the article by Stephen G. Brush, "Should the History of Science Be Rated X?," *Science* 183 (1974): 1164–72. John C. Burnham, *How Superstition Won and Science Lost: Popularizing Science and Health in the United States* (New Brunswick: Rutgers University Press, 1987) gives a detailed account of how the intelligent public's view of science has become less informed and less realistic during this century.

Some acquaintance with the history of science is prerequisite to understanding modern science and popular views of it. Brush provides a marvelous guide in *The History of Modern Science: A Guide to the Second Scientific Revolution, 1800–1950* (Ames: Iowa State University Press, 1988). David Knight gives a very readable appraisal of the culmination in the nineteenth century of the scientific-scientist worldview in *The Age of Science: The Scientific World-view in the Nineteenth Century* (Oxford: Basil Blackwell, 1986).

A concise description of contemporary scientific life, with a select bibliography, by the Committee on the Conduct of Science has been published by the National Academy of Sciences: *On Being a Scientist* (Washington, D.C.: NAS, 1989). How the scientific community manages the manifold details and aspects of enlarging understanding, and the wider social ramifications, is most fully described by John Ziman in *Public Knowledge: An Essay concerning the Social Dimensions of Science* (Cambridge: Cambridge University Press, 1968), *The Force of Knowledge: The Scientific Dimension of Society* (Cambridge: Cambridge University Press, 1976), and *Reliable Knowledge: An Exploration of the Grounds for Belief in Science* (Cambridge: Cambridge University Press, 1978). Ziman has also provided an introduction to the interdisciplinary field of science and technology studies with references for further reading in the individual disciplines of sociology, philosophy, and so on: *An Introduction to Science Studies: The Philosophical and Social Aspects of Science and Technology* (Cambridge: Cambridge University Press, 1986).

Science since Babylon, enlarged ed. (New Haven: Yale University Press, 1975), and *Little Science, Big Science . . . and Beyond,* enlarged ed. (New York: Columbia University Press, 1986), by Derek de Solla Price, a founding father of STS who has made seminal contributions to scientometrics, citation analysis, and technology studies in particular, remain stimulating reading.

David L. Hull, *Science as a Process: An Evolutionary Account of the Social and Conceptual Development of Science* (Chicago: University of Chicago Press, 1988), ventures a massive case study and analysis of the development of scientific knowledge at the hands of competitive human beings, a work that bears comparison with Kuhn's classic on several counts: in appealing more to practicing scientists than to philosophers, in arousing ire among those who (mis)perceive it as ignoring or discarding the rational side of science, and in being influential albeit controversial.

Index

A Note on the Author

HENRY H. BAUER is professor of chemistry and science studies at Virginia Polytechnic Institute and State University, where earlier he served for eight years as dean of the College of Arts and Sciences. A native of Vienna, Austria, Bauer has also held positions at the universities of Sydney (Australia), Michigan, Southampton (England), and Kentucky. His primary area of research has been chemistry, on which he has published several books and more than a hundred articles. More recently, however, he has focused most of his attention on the interaction between science and culture. Bauer is the author of *Beyond Velikovsky: The History of a Public Controversy* (1984), *The Enigma of Loch Ness: Making Sense of a Mystery* (1986), and (under the pseudonym Josef Martin) *To Rise above Principle: The Memoirs of an Unreconstructed Dean* (1988).